PROVING AND PRICING CONSTRUCTION CLAIMS

1992 Cumulative Supplement
Current through August 1, 1991

MARK S. RHODES
Member of the
Illinois and New York Bars

*Insert in the pocket
in the back inside cover
of the bound volume.
Discard supplement
dated 1991.*

Supplement to
Proving and Pricing Construction Claims,
edited by
Robert F. Cushman

Wiley Law Publications
JOHN WILEY & SONS, INC.
New York · Chichester · Brisbane · Toronto · Singapore

ISBN 0-471-55397-2

In recognition of the importance of preserving what has been
written, it is a policy of John Wiley & Sons, Inc., to have books
of enduring value published in the United States printed on
acid-free paper, and we exert our best efforts to that end.

Copyright © 1991, 1992 by John Wiley & Sons, Inc.

All rights reserved. Published simultaneously in Canada.

Reproduction or translation of any part of this work
beyond that permitted by Section 107 or 108 of the
1976 United States Copyright Act without the permission
of the copyright owner is unlawful. Requests for
permission or further information should be addressed to
the Permissions Department, John Wiley & Sons, Inc.

This publication is designed to provide accurate and
authoritative information in regard to the subject
matter covered. It is sold with the understanding that
the publisher is not engaged in rendering legal, accounting,
or other professional services. If legal advice or other expert
assistance is required, the services of a competent
professional person should be sought. *From a Declaration
of Principles jointly adopted by a Committee of the
American Bar Association and a Committee of Publishers.*

Library of Congress Cataloging-in-Publication Data

Proving and pricing construction claims / Robert F. Cushman, David A.
  Carpenter, editors.
      p.   cm. — (Construction law library)
  ISBN 0-471-50913-2. — ISBN 0-471-52724-6 (custom bdg.)
  ISBN 0-471-55397-2 (Supplement)
    1. Construction contracts — United States. 2. Breach of contract —
United States. 3. Construction industry — Law and legislation —
United States. 4. Damages — United States. I. Cushman, Robert
Frank, 1931– . II. Carpenter, David A. (David Allen), 1951– .
III. Series.
KF902.P76   1990
343.73'07869—dc20
[347.3037869]   90-12004
                                                      CIP

Printed in the United States of America

10 9 8 7 6 5 4 3 2 1

PREFACE

This 1992 supplement reflects an increasing emphasis on a wide range of damage issues which are of continuing importance. Specifically, this supplement continues the extensive treatment of the following critical topics.

Additional Work/Change Orders. The supplement contains numerous new cases dealing with the attempts of contractors to recover additional sums in excess of the contract amount. The cases discuss the formal requirements for change orders and the interpretation of what constitutes extra work that will support an award of additional compensation.

Delay Damages. Numerous cases are discussed with regard to a variety of delay claims and whether the delays are attributable to the owner, the contractor, or other causes. There are new cases discussing the problem of liquidated damage provisions contained in construction contracts.

Federal Claims Procedure. This supplement includes a variety of recent cases dealing with the statutory requirements of the Contract Disputes Act. Specifically, there is an emphasis on the claim certification provisions and their effect on jurisdiction.

Causes of Action and Possible Recoveries. The leading chapter requiring supplementation is **Chapter 13**, due to the judicial activity in the area of contractor, subcontractor, architect, and owner liabilities. The supplement again provides extensive coverage of developments in the areas of defective performance and the measure of damages.

Materials that are new to this supplement are indicated by an asterisk (*) in the left margin throughout the supplement.

Rochester, New York
November 1991

MARK S. RHODES

THE CONSTRUCTION LAW LIBRARY FROM WILEY LAW PUBLICATIONS

ALTERNATIVE CLAUSES TO STANDARD CONSTRUCTION CONTRACTS
    James E. Stephenson, Editor

ALTERNATIVE DISPUTE RESOLUTION IN THE CONSTRUCTION INDUSTRY
    Robert F. Cushman, G. Christian Hedemann, and Avram S. Tucker, Editors

ARBITRATION OF CONSTRUCTION DISPUTES
    Michael T. Callahan, Barry B. Bramble, and Paul M. Lurie

ARCHITECT AND ENGINEER LIABILITY: CLAIMS AGAINST DESIGN PROFESSIONALS
    Robert F. Cushman and Thomas G. Bottum, Editors

CONDOMINIUM AND HOMEOWNER ASSOCIATION LITIGATION
    Wayne S. Hyatt and Philip S. Downer, Editors

CONSTRUCTION ACCIDENT PLEADING AND PRACTICE
    Turner W. Branch, Editor

CONSTRUCTION AND ENVIRONMENTAL INSURANCE CASE DIGESTS
    Wiley Law Publications, Editors

CONSTRUCTION BIDDING LAW
    Robert F. Cushman and William J. Doyle, Editors

CONSTRUCTION CLAIMS AND LIABILITY
    Michael S. Simon

CONSTRUCTION DEFAULTS: RIGHTS, DUTIES, AND LIABILITIES
    Robert F. Cushman and Charles A. Meeker, Editors

CONSTRUCTION DELAY CLAIMS
    Barry B. Bramble and Michael T. Callahan

CONSTRUCTION ENGINEERING EVIDENCE
    Loren W. Peters

CONSTRUCTION FAILURES
    Robert F. Cushman, Irvin E. Richter, and Lester E. Rivelis, Editors

CONSTRUCTION INDUSTRY CONTRACTS: LEGAL CITATOR AND CASE DIGEST
    Wiley Law Publications Editorial Staff

CONSTRUCTION INDUSTRY FORMS (TWO VOLUMES)
    Robert F. Cushman and George L. Blick, Editors

CONSTRUCTION LITIGATION FORMBOOK
    David M. Buoncristiani, John D. Carter, and Robert F. Cushman
CONSTRUCTION LITIGATION: REPRESENTING THE CONTRACTOR
    Robert F. Cushman, John D. Carter, and Alan Silverman, Editors
CONSTRUCTION LITIGATION: REPRESENTING THE OWNER (SECOND EDITION)
    Robert F. Cushman, Kenneth M. Cushman, and Stephen B. Cook, Editors
CONSTRUCTION LITIGATION: STRATEGIES AND TECHNIQUES
    Barry B. Bramble and Albert E. Phillips, Editors
CONSTRUCTION RENOVATION FORMBOOK
    Robert F. Cushman and John W. DiNicola, Editors
CONSTRUCTION SCHEDULING: PREPARATION, LIABILITY, AND CLAIMS
    Jon M. Wickwire, Thomas J. Driscoll, and Stephen B. Hurlbut
CONSTRUCTION SUBCONTRACTING: A LEGAL GUIDE FOR INDUSTRY PROFESSIONALS
    Overton A. Currie, Neal J. Sweeney, and Randall F. Hafer, Editors
DESIGN PROFESSIONAL'S HANDBOOK OF BUSINESS AND LAW
    Robert F. Cushman and James C. Dobbs, Editors
DRAFTING CONSTRUCTION CONTRACTS: STRATEGY AND FORMS FOR CONTRACTORS
    Samuel F. Schoninger
FEDERAL CONSTRUCTION CONTRACTING
    James F. Nagle
HANDLING FIDELITY, SURETY, AND FINANCIAL RISK CLAIMS (SECOND EDITION)
    Robert F. Cushman, George L. Blick, and Charles A. Meeker, Editors
HAZARDOUS WASTE DISPOSAL AND UNDERGROUND CONSTRUCTION LAW
    Robert F. Cushman and Bruce W. Ficken, Editors
1990–1991 DIRECTORY OF CONSTRUCTION INDUSTRY CONSULTANTS
    Wiley Law Publications, Editors
1991 WILEY CONSTRUCTION LAW UPDATE
    Steven M. Goldblatt, Editor
PROVING AND PRICING CONSTRUCTION CLAIMS
    Robert F. Cushman and David A. Carpenter, Editors
SWEET ON CONSTRUCTION INDUSTRY CONTRACTS (SECOND EDITION)
    Justin Sweet
TROUBLED CONSTRUCTION LOANS: LAW AND PRACTICE
    Stanley P. Sklar, Editor

CHAPTER 1
WINNING STRATEGIES FOR PROVING AND PRICING CLAIMS

§ 1.3 Defining Rights, Responsibilities, and Risks: The Parties and Their Contracts

Page 9, add at end of section:

Employees of a subcontractor sued the construction manager for injuries they sustained in a fall. In *Wenzel v. Boyles Galvanizing Co.*, 920 F.2d 778 (11th Cir. 1991), applying Florida law, the utility (owner) hired one defendant as the architect and construction manager for the project. As part of the project, the defendant was to:

> Provide, implement and administer a site safety and health program consisting of, but not necessarily limited to, the following elements: (1) Develop a project safety manual that will establish contractor safety guidelines and requirements; (2) Review project contractors' safety programs for compliance with the project safety manual; (3) Provide daily surveillance of contractor work areas for compliance with safety program; (4) Develop and invoke procedures for advising contractors of safety violations and deficiencies; (5) Develop and invoke procedures for initiating corrective action by the commission and backcharge to the contract, if contractor does not comply with safety violation directives.

The defendant had the authority to issue notices of safety violations, remove employees from the job, and stop the work of any particular contractor. However, the contract also provided:

> [T]he Construction Manager shall have no responsibility or right to exercise any actual or direct control over employees of the Contractors. The obligations assumed by the Construction Manager hereunder run to and are for the sole benefit of the [owner].... [T]he furnishing of such services shall not make the Construction Manager responsible for construction means, methods, techniques, work sequences or procedures....

The jury found the construction manager 50 percent at fault for the fall. This was affirmed on appeal, the court holding that the contractual provisions did not relieve the manager of liability in negligence. There was evidence for the jury to determine whether the defendant should have required the placement of safety nets below each floor. The court further found that the manager was not a "safety consultant" for the subcontractor-employer so as to place the injured worker's claim within the exclusive remedy provisions of workers' compensation.

A homeowner sued the contractor following the contractor's failure to complete a home on time. In *Schlothauer v. Gusse*, 753 F. Supp. 414 (D. Mass. 1991), the defendant proposed a contract with an "upset price" of $239,640, which was the maximum price that would be charged unless the owners (plaintiffs) made unusual or excessive changes. The proposal was rejected. The defendant completed another proposal with a lower upset price. If the house was completed for less than the bid, the plaintiffs would receive 70 percent of the savings. If the cost was over the upset price, they would be responsible for unusual charges or excessive changes or amounts in excess of specific allowances. This contract was accepted, with a scheduled November completion date. During construction, the defendant told the plaintiffs that he could complete a better design within the contract price. Various change orders were signed and, during the construction, the plaintiffs agreed to sell the house to another person after being told that the home would not be ready before Christmas. Work was stopped several times due to the defendant's lack of financial resources and ultimately the plaintiffs hired another contractor to complete the house. Due to the delay, the plaintiffs spent an additional $4,000. The court held that the plaintiffs were not entitled to recover for the costs incurred due to the delays where it found that their failure to make the selection of finish goods in a timely manner and their insistence on unnecessary changes caused the delay.

In *Lang Bros. Inc. v. United States*, 20 Cl. Ct. 551 (1990), the plaintiff sought amounts expended for repair of a cofferdam. The plaintiff was experienced in constructing dams and had previously built various dams for the government. The plaintiff contracted to construct an earthen dam in incremental stages over a three-year period. The contractor was to construct a permanent dam, a dike, and a cofferdam. The cofferdam was a temporary measure to rechannel water during construction of the permanent dam and was upstream from the permanent dam. The cofferdam diverted the water into a 66-inch concrete pipe and provided a dry streambed in which the permanent dam could be built. The contract specifically dealt with the risk of damage to the project from flooding during the construction process. The contract provided that the government would assume the risk of damage to the permanent work at a crest of water below a certain elevation. The contract stated:

> If either the dike or cofferdam fail, due to no fault of the Contractor, at or below the above stated elevations, a modification will be issued by the Contracting Officer to provide an equitable adjustment in costs to restore the damaged work.

Problems with the driving of certain pilings rendered the cofferdam vulnerable to flooding. The government reminded the contractor that it would not be responsible for damages due to failure of the pilings. The court held that there were factual questions which precluded a grant of summary judgment with respect to whether the contractor was at fault for the collapse of the cofferdam due to its failure to drive the pilings to refusal.

§ 1.5 STANDARD FORMS, PROVISIONS

In *Mrozik Construction, Inc. v. Lovering Associates, Inc.*, 461 N.W.2d 49 (Minn. Ct. App. 1990), a subcontractor sued to recover amounts due under its subcontract. The general contractor refused to make a payment until it was paid by the owner. The amounts due were never paid to the general contractor because the owner became insolvent. The subcontractor had completely and properly performed under its subcontract. The subcontract provided:

> THE CONTRACTOR AGREES AS FOLLOWS:
> D. Final payment including all retention becomes due and payable within 30 days after Architects' certification of final payment. At all times the Subcontractor shall be paid to the extent that the Contractor has been paid on the Subcontractor's account.

This clause was present in the form published by the Associated General Contractors of Minnesota and was entitled "Standard Subcontract Agreement." The court affirmed the grant of summary judgment for the subcontractor, holding that the language was not sufficient to shift the burden of the owner's nonpayment to the subcontractor. The fact that the architect never certified the project for final payment did not affect the subcontractor's right to be paid under these circumstances.

In *Kinney v. G.W. Lisk Co.*, 76 N.Y.2d 215, 556 N.E.2d 1090, 557 N.Y.S.2d 283 (1990), an employee of a subcontractor was injured on the construction site and sued the general contractor and the owner, whereupon the general contractor brought a third-party claim against the subcontractor. The subcontract provided that the:

> Subcontractor shall maintain such insurance policies . . . as will protect both the Contractor and the Subcontractor . . . from claims for damages because of bodily injury . . . which may arise both out of and during its performance under this Agreement or after completion thereof.

The court upheld this agreement, finding that the contract provision did not violate N.Y. Gen. Oblig. Law § 5-322.1. On this basis, the court affirmed the summary judgment granted to the general contractor on the third-party claim, because the subcontractor breached the agreement. The subcontractor was therefore liable for any resulting damages, including any potential liability imposed on the general contractor for the plaintiff's injuries.

§ 1.5 —Standard Contract Forms and Key Contract Provisions

Page 11, add at end of section:

In *Aetna Casualty & Surety Co. v. Canam Steel Corp.*, 794 P.2d 1077 (Colo. Ct. App. 1990), the building in question collapsed. The insurer, as subrogee of the property owner, sued the architect and the materialman for the resulting loss. The subrogor-owner had contracted with the general contractor for an addition to its church. The contractor used the American Institute of Architects form, which provided that the owner and the general contractor waived all rights against "subcontractors" and against the architect and its agents for "damages caused by fire or other perils to the extent covered by insurance. . . ." The insurer had issued a builders' risk policy to the church and paid $1.4 million for investigation of the claim and reconstruction of the church.

The trial court granted the defendant's motion for summary judgment on the basis that the general contract provided for a waiver of the owner's claim and the insurer's derivative claim. The appellate court reversed, holding that because the defendant materialman did not perform work on the church, it was not a "subcontractor" and therefore was not within the waiver of claims. There was also a factual issue as to the intent of the owner with respect to the waiver of claims against the architect. This precluded the grant of summary judgment.

§ 1.12 —Enforceability of Agreements to Arbitrate

Page 16, add at end of section:

In *Henderson Investment Corp. v. International Fidelity Insurance Co.*, 575 So. 2d 770 (Fla. Dist. Ct. App. 1991), the owner contracted with the general contractor for the construction of a service center. The owner and the contractor agreed in the contract that all successors, assigns, and legal representatives would be bound by the contract, which provided for the arbitration of disputes. The contract, however, also provided that:

> No arbitration shall include by consolidation, joinder or in any other manner, parties other than the owner, the contractor and any other person substantially involved in a common question of fact or law, whose presence is required if complete relief is to be accorded in the arbitration.

The contractor breached the contract and the owner sued the surety. The surety then sought to compel arbitration under the terms of the construction contract. The court held that the surety's obligations were coextensive with those of its principal contractor and, because the bond incorporated the terms of the contract, the surety could compel arbitration of the dispute.

In *D.M. Ward Construction Co. v. Electric Corp.*, 15 Kan. App. 2d 114, 803 P.2d 593 (1990), an electrical subcontractor sued the general contractor to recover the balance due under a subcontract. The general contractor contracted with the owner to build a warehouse-distribution center. The plaintiff subcontractor was selected as electrical contractor and a written subcontract was signed. The American Institute of Architects standard form subcontract was used and provided: "All claims, disputes and other matters in question arising out of, or relating to, this subcontract . . . shall be decided by arbitration." The subcontractor was asked to perform additional work not contained in the subcontract, and there was a dispute as to whether still other work was within the subcontract. The subcontractor completed its work but was not paid the full contract price nor for the extra work

§ 1.12 ENFORCEABILITY OF ARBITRATION

it performed. The subcontractor filed a mechanic's lien against the property, after which it received a partial payment. Suit was then filed.

The court held, respecting the arbitration provisions, that the general contractor had waived the right to demand arbitration when its actions were inconsistent with the right to arbitration, in that the general contractor failed to file a timely motion to compel after the suit was first filed. With respect to the interpretation of whether the extra work was outside the scope of the subcontract, the court held for the subcontractor, affirming the judgment of nearly $24,000.

In *Crown Oil & Wax Co. v. Glen Construction Co.*, 320 Md. 546, 578 A.2d 1184 (1990), the owner's assignee sued to compel the contractor to proceed to arbitration. The question was whether the contractor was required to arbitrate claims made by a limited partnership made up of the same individuals who controlled the owner of the property. The limited partnership was used to syndicate the project, which was for the construction of a Quality Inn for the franchisee owner. The construction contracts used the American Institute of Architects Document No. A201 (1976 ed.). The court held that the assignee was entitled to compel the arbitration. In order to compel the arbitration, the court may not inquire as to the merits of the claims. Rather, the court is only to determine whether the demand falls within the arbitration provisions of the contract. This contract contained a broad arbitration clause which could be invoked where the plaintiff was the equitable assignee of the benefits under the construction contract and where the assignee assumed the obligations of the owner. The court clearly distinguished this situation from any third-party beneficiary theory.

In *North West Michigan Construction, Inc. v. Stroud*, 185 Mich. App. 649, 462 N.W.2d 804 (1990), the homeowners were found to have waived their right to arbitration of a dispute with the contractor. The construction contract provided that any controversies were to be resolved by arbitration. The defendant homeowners filed their answer to the contractor's complaint in June 1988 but did not move to dismiss the complaint on the basis of the arbitration provisions until more than eight months had passed. The court was required to determine whether this conduct constituted a waiver of the right to invoke arbitration. The court affirmed the judgment for the contractor, holding that the homeowners had waived the right to arbitration which would have provided the basis for a motion to dismiss the suit. The court construed the contract as requiring that the contractor complete the house within 90 days of obtaining necessary permits and not within 90 days of the date of execution of the contract, as the homeowners had contended.

CHAPTER 5

DELAY CLAIMS

§ 5.2 Delays and Cause of Delays

Page 102, add to footnote 1:

   In connection with a lock construction project on the Tennessee-Tombigbe Waterway, the general contractor subcontracted with an electrical subcontractor, which in turn contracted with a supplier for various control and electrical power systems. Subsequently, a substantial change order was executed by the Army Corps of Engineers and delivered to the electrical subcontractor, who failed to transmit the information to the supplier. This failure resulted in a delay in the supplier's completion of its work and increased expenses on the part of the general contractor. The general contractor withheld funds from the subcontractor, claiming that the subcontractor was at fault for not timely transmitting the change to the supplier. The Fifth Circuit held that the general contractor was not a third-party beneficiary of the contract between the subcontractor and the supplier, and therefore the general contractor could not recover for the supplier's late performance. The supplier could not be held liable in contract or in negligence for the delay. United States *ex rel.* Control Systems, Inc. v. Arundel Corp., 896 F.2d 143 (5th Cir. 1990) (applying Mississippi law).

* A contractor agreed to perform construction work at six public schools. The plaintiff contractor sought to obtain additional amounts in excess of the contract price as delay damages, after encountering delays at all construction sites. The board of education awarded an additional $145,000 in compensation on the contracts and an additional amount for delay damages on a seventh project which was subsequently started. The authorizations for the additional payments were never acted upon by all of the various authorities required to approve the payment. The court affirmed dismissal of the contractor's action in Manshul Constr. Corp. v. Board of Educ., 160 A.D.2d 643, 559 N.Y.S.2d 260 (1990). The court held that even though the construction contract contained a no damage for delay exculpatory clause, the plaintiff had the burden of proving that the delays experienced were wholly unanticipated. In this case, the mere evidence of additional costs due to delay did not meet this burden.

* In State Highway Admin. v. Greiner Eng'g Sciences, Inc., 83 Md. App. 621, 577 A.2d 363 (1990), a contractor sued to recover delay damages. The delays occurred during the plaintiff's preparation of contract documents for a highway project. The plaintiff agreed to perform final design services for the project, which included the final bridge and road design and preparation of the contract

§ 5.2 DELAYS AND CAUSE

plans, specifications, and documents for bid advertisement. The plaintiff was also to review shop drawings and any construction redesigns. The payment was to be on a cost plus fixed fee basis subject to an overall maximum amount. The defendant rewrote the contract to include a no damages for delay clause which provided:

> The Consultant agrees to prosecute the work continuously and diligently and no charges or claims for damages shall be made by him for any delays or hindrances, from any cause whatsoever during the progress of any portion of the services specified in this Agreement. Such delays or hindrances, if any, may be compensated for by an extension of time for such reasonable period as the Department may decide. Time extensions will be granted only for excusable delays such as delays beyond the control and without the fault or negligence of the consultant.

The court held that the clause was enforceable even though the delay was caused by funding problems that were outside the contemplation of the parties. The court held that the clause was not unconscionable. The problems extended the duration of the work from 15 months to more than 6 years.

*   In White Oak Corp. v. Department of Transp., 217 Conn. 281, 535 A.2d 1199 (1991), a contractor sued the department to recover the balance due under a contract for highway construction. Specifically, the contractor sought to recover its delay damages. The court was required to interpret a no damages for delay clause. The contract required that the contractor complete the work within 1,650 days from the starting date. If the contractor failed to complete the project within that time, the contract permitted the department to deduct $1,000 per day for each day after the specified completion date. The contractor completed the work 392 days late. After completion the contractor invoked a contract provision which permitted it to seek a retroactive extension of the completion date due to delays beyond the contractor's control. The department extended the date by 305 days for excusable delays and "winter shutdown." This resulted in a payment deduction by the department of $87,000. In arbitration, the decision was that all the delay was caused by factors outside the contractor's control, such as the failure of a gas company to remove its gas lines and the department's own delay in furnishing controls for the construction of temporary railroad lines. The arbitration found that the delays also resulted in other damages to the contractor for increased labor and overhead costs. The arbitration also found that the department failed to make timely periodic payments pursuant to the contract and that payments were required for extra work and departmental overcharges to the contractor for gravel the contractor removed during the project. The trial court entered judgment for the contractor on most of the arbitration award. The appellate court reversed in part, holding that the contractor could not recover delay damages attributable to the failure of the gas company to perform in a timely manner. The court held that the no damages for delay clause was not enforceable with respect to claims for bad faith, willful or grossly negligent conduct, uncontemplated or unreasonable delay, or from delays due to one party's breach of a fundamental contractual undertaking. The court also awarded the contractor interest on the late periodic payments.

DELAY CLAIMS

Page 102, add to footnote 2:

In Fridlen v. Winchendon Hous. Auth., 28 Mass. App. Ct. 977, 553 N.E.2d 554 (1990), a contractor sued a housing authority for delay damages. The contractor rejected a low bid for roofing, due to prior experience with the subcontractor, and proposed that it perform the roofing work itself. This was approved by the housing authority after a delay of approximately seven months. Meanwhile, the lack of cover on the buildings disturbed the scheduling of other subcontractors. These subcontractors' delay damages were the subject of this suit, as the plaintiff contractor had already been determined to be liable for such damages in a prior action. The contract between the contractor and the authority provided: "Apart from extension of time for unavoidable delays, no payments or allowances of any kind shall be made to the Contractor as compensation for damages because of hindrance, or delays from any cause in the progress of work whether such delay be avoidable or unavoidable." 553 N.E.2d at 556. The appeals court affirmed the lower court's award of delay damages to the contractor, holding that the conduct of the authority waived the authority's right to rely on the above-quoted contract provision.

Page 102, add to footnote 3:

In Stone v. City of Arcola, 181 Ill. App. 3d 513, 536 N.E.2d 1329 (1989), the plaintiff contractor sued the city over work performed in connection with the construction of a new sanitary sewage facility. The contractor was to start work within 10 days of the notice to proceed, and completion was to be within one year unless otherwise extended. The notice to proceed was issued following the submission of payment and performance bonds. The notice to proceed contained a completion date 10 days after that originally stated. The contractor sought an extension due to abnormal amounts of rainfall. In order to accommodate a request of the city, the contractor performed some work out of sequence; there were also other disputes over the adequacy of the plans and specifications. These problems were corrected by engineers in the field, but resulted in additional delays. With the scheduled completion date nearing, the contractor sought a 90-day extension, but was only granted 60 days and was reminded that the city would enforce the liquidated damage provision of $200 per day for delays beyond the extension. Under these facts and conflicting expert testimony, the trial court entered judgment for the contractor.

The appellate court affirmed the judgment, holding that the liquidated damage provisions were reasonable and generally enforceable. The running of the delay period, under that provision, terminated when the project was 95% completed, with only minor repairs and finishing work remaining to be done. The contractor was entitled to recover damages in the form of increased costs sustained due to delays that were attributable to the city. The court, however, rejected the city's counterclaim for damages arising out of the delayed completion of the project, since the city's actions caused the delay.

§ 5.2 DELAYS AND CAUSE

Fred A. Arnold, Inc. v. United States, 18 Cl. Ct. 1 (1989), was an appeal by both parties from a decision of the Armed Services Board of Contract Appeals arising out of the construction of military housing units. The contractor was seeking damages for goverment-caused delays and disruptions and for extra work. The government sought liquidated damages for the 13-month delay in project completion. The initial construction was completed in a timely manner, but the balconies required structural modifications which delayed the use of the housing units for 13 months. The board of appeals rejected almost all of the contractor's claims and denied the government's claim for liquidated damages. There were substantial negotiations and a settlement was arranged. The board of appeals held that the settlement worked an accord and satisfaction, but the contractor claimed that the consideration was inadequate for such a settlement. The court held that there was a valid accord and satisfaction that resolved most of the claims. The court also found that the evidence supported the finding that the Navy did not cause unnecessary delays, and awarded liquidated damages pursuant to the contract provision, consistent with government regulations. Such liquidated damages were recoverable without proof that actual damages were suffered.

Page 103, add to footnote 5:

In North Star Contracting v. Long Island R.R., 723 F. Supp. 902 (E.D.N.Y. 1989), the contractor sued the defendant railroad for the wrongful termination of its contract. The contractor agreed to perform construction work at the railroad's support facility, but the defendant railroad caused numerous delays, and there were defective contract documents. The defendant also failed to provide sufficient insurance, and there were other claims which could be characterized as sounding in fraud and mismanagement. The court, however, dismissed claims against the railroad for violations of RICO and for the deprivation of due process and equal protection rights. With respect to the claims against the railroad, all claims excepting those for breach of contract were dismissed. The defendant was a subsidiary of the metropolitan transit authority and had no immunity for a breach of contract claim.

* *Page 103, add at end of section:*

A contractor's default resulted in damages sustained by the plaintiff materialman in *Salvino Steel & Iron Works, Inc. v. Fletcher & Sons, Inc.*, 398 Pa. Super. 86, 580 A.2d 853 (1990). The statutory surety bond provided:

> The principal and Surety hereby jointly and severally agree with the obligee herein that every person, copartnership, association or corporation who, whether as Subcontractor or as a person otherwise entitled to the benefits of this Bond, has furnished material or supplied or performed labor or rented equipment used in the prosecution of the work as above provided . . . and who has not been paid in full therefor, may sue in assumpsit on this Bond . . . and prosecute the same to final judgment. . . .

The contractor voluntarily defaulted on the contract and the surety negotiated a new contract with another contractor. The new contractor subcontracted with the plaintiff to perform the same work as provided for in the original agreement. The delay caused by the initial contractor's default resulted in the plaintiff's incurring additional cost in storing steel and renting trailers for the storage during a four-month period. This was the basis for the plaintiff's claim against the bond. The court affirmed the summary judgment for the surety, holding that there was no liability under the bond where the bond did not specifically provide coverage for delay damages such as those claimed.

A prime contractor's Miller Act surety was sued in *United States ex rel. Pertun Construction Co. v. Harvesters Group*, 918 F.2d 915 (11th Cir. 1990) (applying Miller Act), wherein a subcontractor sought additional amounts to cover increased costs attributable to delays caused by the prime contractor. The issue of whether the subcontractor was entitled to recover from the Miller Act surety was a question of first impression in the Eleventh Circuit. There was also a question as to whether, under the terms of the subcontract, the sole remedy of the subcontractor in such a situation would be an extension of the time for performance. The contractor secured a payment bond pursuant to the Miller Act and then contracted with the subcontractor to perform concrete work for the project. Toxic wastes were discovered on the job site, the site lacked running water and electricity, the contractor had difficulty finding subcontractors to perform certain parts of the work, and the contractor was unable to obtain necessary documents from the government in a timely manner. All this resulted in significant construction delays. The prime contractor also failed to supervise the project and properly coordinate the work of the subcontractors. This resulted in further delays for the subcontractor, which was subsequently and wrongfully terminated after it had completed 80 percent of its work. The subcontractor was not permitted to complete its performance or to retrieve its tools and materials. The court held that the surety was liable for the increased expenses sustained by the subcontractor due to the prime contractor's actions, under the bond provision covering any "sum or sums justly due" for labor or materials. The court further held that the subcontractor's remedy was not limited by the terms of the subcontract when the contract's conditions precedent to the exclusive remedy were never fulfilled.

§ 5.3   Excusable and Nonexcusable Delay

Page 104, add to footnote 8:

Delay damages were sought by a subcontractor from the general contractor and its surety in Port Chester Elec. Constr. Corp. v. HBE Corp., 894 F.2d 47 (2d Cir. 1990) (applying New York law). The plaintiff was an electrical subcontractor involved in the renovation of a hospital. The plaintiff sought additional sums arising out of a number of design changes, errors, stop-work orders, and other actions that increased the cost of performing the job. The contract provided:

§ 5.3 EXCUSABLE/NONEXCUSABLE

"Contractor may, from time to time, modify or alter the work schedule, but in such an event, no such modification or alteration shall entitle Subcontractor to any increase in the consideration of the Subcontract." 894 F.2d at 48. The Second Circuit found this contract language to be ambiguous and reversed the trial court's dismissal of the suit. The court based its decision on the general precept that clauses seeking to exempt one from the consequences of one's own fault are to be strictly construed. The court found that the provision was not sufficiently clear and unambiguous so as to be enforceable.

In Dale R. Horning Co. v. Falconer Glass Indus. Inc., 730 F. Supp. 962 (S.D. Ind. 1990), the plaintiff was a glazing subcontractor who installed a glass curtain wall in a construction project. The plaintiff was operating under substantial time constraints. The supplier of the glass was sued by the subcontractor for supplying defective glass. The defendant supplier knew of the particular requirements for the glass in question. The defendant's standard form provided that the glass would be free from defects, but also provided that the buyer's "exclusive and sole remedy" for defective goods "shall be to secure replacement." The defendant would not be liable for "special, direct, indirect, incidental or consequential damages." These provisions were in fine print and, under the UCC, did not become part of the contract. The plaintiff was entitled to recover consequential damages sustained due to the defective glass supplied by the defendant.

Gymco Constr. Co. v. Architectural Glass & Windows, Inc., 884 F.2d 1362 (11th Cir. 1989) (applying Georgia law), arose out of the construction of an exercise salon. A mirrored glass facade was to be installed. The contract specified a completion date following the receipt of approved shop drawings and the delivery of materials. A template was to be provided for the location of holes to be drilled for the anchoring of a sign. The template was to be created by another subcontractor, the sign company. The template arrived five weeks late, which did not permit the installation of the glass by the contract completion date.

A salesman for the contractor made an oral agreement to switch from the glass to stainless steel. Upon the delivery of the steel, the contractor refused to install the steel, which was more expensive than the glass and was outside the company's expertise. Any such agreement between the general contractor and the subcontractor's salesperson was clearly outside the scope of the salesperson's authority. The general contractor then contracted with another subcontractor to supply and install the steel panels. The price of the steel and its installation was $50,414 higher than for the glass. The general contractor then sued the glass subcontractor for the difference, in a breach of contract action.

The subcontractor claimed that the general contractor had interfered with its performance of the contract. The court held that the glass subcontractor did not breach the contract by failing to install the glass in a timely fashion, since it did not receive the template on schedule. This delay constituted a breach of contract by the general contractor. The fact that the subcontractor continued to perform other work on the project constituted a waiver of the subcontractor's right to terminate the contract for the general contractor's breach. The subcontractor, however, retained its claims for damages caused by the delay in furnishing the

DELAY CLAIMS

template. Therefore, the court reversed the trial court's award of damages to the general contractor, and the case was remanded.

* In Blau Mechanical Corp. v. City of N.Y., 158 A.D.2d 373, 551 N.Y.S.2d 228 (1990), the plaintiff contracted to do the plumbing work for a simulated tropical rain forest in a public park. The plaintiff sought damages due to delays allegedly caused by changes in the structure of the exhibit, which were required to be approved by various municipal departments. Some change orders called for additional excavations due to subsurface conditions that differed from the city's plans. Other delays were caused when members of the public demonstrated on the job site and threatened the workers, thereby temporarily closing down the project. The court reversed the denial of the city's motion for summary judgment, holding that the delays due to the additional excavations could not support a claim for delay damages since the information given to all bidders stated that contractors could request a modification of the contract if unusual subsurface conditions were discovered that would materially affect the cost of work. Therefore, the potential delays in such a contingency were anticipated and the plaintiff was awarded a 25% increase in the contract price due to the additional work. Similarly, the city had contractually reserved the right to make such changes in the design and structure of the exhibit. As for the delay caused by the public protesters, there was no evidence that any such delay was attributable to the city and therefore no recovery was available for that delay claim.

§ 5.6 Burden of Proof and Apportionment in Delay Cases

Page 108, add to footnote 25:

A subcontractor sued the prime contractor and recovered under its subcontract for labor and materials. The subcontract specifically provided that the prevailing party in any such breach of contract litigation was entitled to recover attorneys' fees. The claim involved delay damages sustained by the subcontractor arising from a 15-month delay caused in large part by the owner. The case of David C. Olson Co. v. Denver & Rio Grande W. R.R., 789 P.2d 492 (Colo. Ct. App. 1990), involved the prime contractor's attempt to recover the amounts it paid to the subcontractor from the owner. The prime contractor's contract with the owner did not provide for the recovery of delay damages. The court affirmed the lower court judgment, holding the owner railroad liable for a share of the attorneys' fees paid to the subcontractor in an amount related to its degree of fault for the delay and additional costs. The court imposed this liability despite the fact that the owner was not a party to the subcontract.

* In Allen & O'Hara, Inc. v. Barrett Wrecking, Inc., 898 F.2d 512 (7th Cir. 1990) (applying Wisconsin law), the plaintiff general contractor hired the defendant to perform demolition work on a portion of an office building as part of a renovation project. The contract provided a schedule for demolition that could not be accomplished because the owner delayed three and one-half months in vacating the building. The defendant then submitted a four-month completion

§ 5.6 PROOF AND APPORTIONMENT

schedule and the contract was amended to provide for the work to be performed over that period. The defendant encountered unanticipated conditions which resulted in more delays. The general contractor also required that the defendant use various procedures to reduce dust, noise, and vibrations. The defendant submitted a new completion schedule but, because of the delays, the plaintiff terminated the contract. Thereupon, the plaintiff sued the demolition contractor and its surety for breach of contract. The issue revolved around which party was responsible for the delays and which party should bear the cost of additional demolition. The court held that there was evidence to support a finding that the contract had been modified to cover the extra work claimed by the demolition contractor. The court found that the subcontract provisions requiring that change orders be in writing had been waived by the parties.

* Lakeview Constr. Co. v. United States, 21 Cl. Ct. 269 (1990), was a suit by a contractor seeking to recover impact and delay costs occasioned by change orders. During the course of performance of the contract, the Navy sought certain changes. The contractor reserved its right to claim impact and delay costs for each of the 22 modifications made pursuant to change orders. The question was whether the various paperwork submissions complied with the requirements of the Contract Disputes Act (41 U.S.C. § 601 *et seq.*). The contractor's paperwork consisted of various letters with attachments signed by officers of the contractor and its counsel, some of which did not specifically demand a decision on the claim by the contracting officer. The court set out the language of various letters, some of which did request a final decision by the contracting officer. The court, however, granted the government's motion to dismiss the complaint due to the failure of the contractor to submit the claims to the contracting officer. Instead, the contractor had incorrectly presented the claim to the resident officer in charge of construction. This was not in compliance with the notice requirements of the statute. This case points out the need for strict compliance with the statutory requirements.

* In Walsky Constr. Co. v. United States, 20 Cl. Ct. 317 (1990), a contractor sought to recover delay damages sustained while repairing a hangar at an Air Force base. Modifications of the contract extended the completion date, but the plaintiff did not complete the work within the extension and was assessed liquidated damages. On the contractual completion date, the plaintiff submitted a claim alleging that the Air Force supplied defective specifications, made numerous design changes during the course of construction, and failed to respond timely to job-site problems. The contractor sought to compel discovery of two documents. One was an investigative report by the inspector general and the other was a report by the Office of Special Investigation. The first report evaluated the Air Force's contract administration and the second report was an investigation of this particular contract and its administration. The defendant objected to the disclosure on the grounds of executive privilege. The court granted the plaintiff's motion for discovery of the first report and ordered it to be produced. The court rejected the privilege arguments with respect to that report because the report did not contain predecisional information and did not contain opinions or recommendations that would be covered by the privilege. The second report, however, did contain such materials and was thus protected.

* In Century Constr. Co. v. United States, 22 Cl. Ct. 63 (1990), a contractor sought an equitable adjustment of the contract price, on its behalf and on behalf of its subcontractors, arising from delays caused by defective contract specifications. The plaintiff also alleged that the government failed to cooperate with it. The Corps of Engineers awarded the plaintiff a contract for construction of a new medical and dental clinic at an Air Force base. The plaintiff encountered difficulties due to missing and incorrect dimensions in the contract plans. This caused discrepancies between interior partition walls, doorways, and structural members, and problems with mechanical and electrical systems. The plaintiff notified the government of the problems through "Request for Information" letters. The plaintiff submitted a claim to the contracting officer, on behalf of its subcontractors, for an equitable adjustment of the contract price. The claim stated:

> Century Construction hereby certifies that the foregoing claim is made in good faith, that the supporting data are accurate and complete to the best of our knowledge and belief, and that the amount of the claim requested accurately reflects the contract adjustment for which the *subcontractor* believes the Government is liable [emphasis added].

The certification was signed by the president of the plaintiff. There were separate certifications for the plaintiff's own claims, using identical language. The contracting officer denied the claims. The court granted the government's motion for a partial dismissal, holding that the certification for the subcontractor failed to meet the requirements of the Contract Disputes Act. Specifically, the contractor failed to certify that the amount claimed accurately reflected the amount for which the *contractor* believed the government was liable. The last sentence of the quoted portion of the claim was statutorily deficient.

* In Sorrels Steel Co. v. Great S.W. Corp., 906 F.2d 158 (5th Cir. 1990) (applying Florida law), a subcontractor that worked on a city arts center construction project sued the general contractor. The subcontractor was to detail, fabricate, and deliver structural and other steel. The subcontract was in the form of a purchase order which provided for inspection of the fabricated steel. The purchase order explicitly excluded testing. The subcontractor was not able to maintain a continuous fabrication schedule because of incomplete and inaccurate contract drawings, slow approval of shop drawings, and numerous design changes. The city then required a thorough testing program. The subcontractor agreed to the testing but sought a change order since testing had been specifically excluded from the contract and increased the production time. The subcontractor sought additional compensation for the delays and extra work. The general contractor forwarded the claims to the city and told the subcontractor that it would be advised of the progress of the claims. After the claims were not satisfactorily resolved, the subcontractor filed suit against the general contractor. The general contended that the damages sustained by the sub were due to the actions of the city and its architect and did not arise

§ 5.7 LIQUIDATED DAMAGES CASES

from a breach of the purchase order. The court held that the general contractor was not liable for damages sustained by the subcontractor prior to an addendum to the purchase order, which was designed to resolve the problem of the design changes and improper drawings. After the addendum, the general could be held liable for resulting damages. The court also held that the general contractor could be held liable for damages attributable to the testing. This was proper despite the fact that the city never issued a change order for the testing, as the court interpreted the addendum to the purchase as an assurance by the general contractor that such a change order would be issued.

* In Giuliani Contracting Co. v. United States, 21 Cl. Ct. 81 (1990), the plaintiff contracted with the Army to replace sanitary sewers at West Point. During performance, the contractor claimed that it sustained delay damages. There were work suspensions, differing site conditions, and design defects which allegedly delayed the contractor's performance. The plaintiff submitted over 30 claims to the contracting officer. The contracting officer determined that three audits did not support the delay claims. Finally, the contracting officer denied 20 unresolved claims and the plaintiff appealed this decision to the court. The government sought to transfer the suit to the Armed Services Board of Contract Appeals, where two other claims were still pending. The Contract Disputes Act provides:

> If one or more suits arising from one contract are filed in the United States Claims Court and one or more agency boards, for the convenience of the parties or witnesses or in the interest of justice, the United States Claims Court may order the consolidation of such suits in that court or transfer any suits to or among the agency boards involved.

In this case, the court chose to grant the government's motion to transfer the action to the Armed Services Board of Contract Appeals.

§ 5.7 Apportionment in Liquidated Damages Cases

Page 110, add to footnote 38:

In City of Elmira v. Larry Walter, Inc., 150 A.D.2d 129, 546 N.Y.S.2d 183 (1989), a contractor walked off the job and was sued by the city. The construction contract provided for liquidated damages in the amount of $1,000 per day for each day the project was delayed beyond the scheduled completion date. The contract also provided that for each day the project was completed ahead of schedule, the contractor would receive an additional $1,000. The contractor was to receive progress payments certified by the consulting engineer. The engineer authorized the first four such payments, approved the fifth payment in part, and rejected the entire claim for the sixth, seventh, and eighth payments because the city believed it was being overcharged. There were also other outstanding issues involving changes in the plans. The contractor stopped work. The court recognized the enforceability of liquidated damage clauses, but held that such provisions

DELAY CLAIMS

were inoperative where the contractor abandoned the work prior to the scheduled date for completion. The court held that the city was entitled to recover the cost of completing the parking garage, less any amounts not already paid under the original contract.

§ 5.8 Computing the Delay

Page 113, add to footnote 46:

In Stone v. City of Arcola, 181 Ill. App. 3d 513, 536 N.E.2d 1329 (1989), the plaintiff contractor sued the city over work performed in connection with the construction of a new sanitary sewage facility. The contractor was to start work within 10 days of the notice to proceed, and completion was to be within one year unless otherwise extended. The notice to proceed was issued following the submission of payment and performance bonds. The notice to proceed contained a completion date 10 days after that originally stated. The contractor sought an extension due to abnormal amounts of rainfall. In order to accommodate a request of the city, the contractor performed some work out of sequence; there were also other disputes over the adequacy of the plans and specifications. These problems were corrected by engineers in the field, but resulted in additional delays. With the scheduled completion date nearing, the contractor sought a 90-day extension, but was only granted 60 days and was reminded that the city would enforce the liquidated damage provision of $200 per day for delays beyond the extension. Under these facts and conflicting expert testimony, the trial court entered judgment for the contractor.

The appellate court affirmed the judgment, holding that the liquidated damage provisions were reasonable and generally enforceable. The running of the delay period under that provision terminated when the project was 95% completed, with only minor repairs and finishing work remaining to be done. The contractor was entitled to recover damages in the form of increased costs sustained due to the delays that were attributable to the city. The court, however, rejected the city's counterclaim for damages arising out of the delayed completion of the project, since the city's actions caused the delay.

§ 5.10 Pricing the Claim

Page 120, add at end of section:

The amount recoverable as overhead under a change order, as provided for in the contract, was the subject of *Reliance Insurance Co. v. United States*, 20 Cl. Ct. 715 (1990). The contractor sought additional compensation in connection with its delay claim incurred during the building of an addition to a Veterans Administration hospital. During the course of the project, six separate claims arose. The contract provided:

§ 5.10 PRICING THE CLAIM

> Allowances not to exceed 10 percent each for overhead and profit for the party performing the work will be based upon the value of labor, material and use of construction equipment required to accomplish the change. As the value of the change increases, a declining scale will be used in negotiating the percentage of overhead and profit. . . . Profit shall be computed by multiplying the profit percentage by the sum of the direct costs and computed overhead costs.

Overhead was defined as:

> Overhead and Contractor's fee percentages shall be considered to include insurance other than mentioned herein, field and office supervisors and assistants, watchman, use of small tools, incidental job burdens, and general home office expenses, and no separate allowances will be made therefore [sic]. Assistants to office supervisors includes all clerical, stenographic and general office help. Incidental job burdens include, but are not necessarily limited to, review and coordination, estimating and expediting relative to contract changes are associated with field and office supervision and are consideered to be included in the contractor's overhead and/or fee percentage.

The court held that the contractual limitations on the recovery of overhead for change orders applied equally to claims for delay damages and for direct costs. Therefore, the court granted the government's motion for partial summary judgment.

CHAPTER 6
DISRUPTION CLAIMS

§ 6.2   Delay versus Disruption

Page 145, add at end of section:

In *Mor-Wood Contractors, Inc. v. Ottinger*, 205 Ill. App. 3d 132, 562 N.E.2d 1247 (1990), the general contractor sued the homeowners to recover for services rendered in connection with construction of a house, which was only partially completed. The homeowners purchased an undeveloped parcel of property, filed a site plan for a proposed septic field and drainage swale, and hired an architect to draw up plans for a house. The parties negotiated for construction of the house. The owners gave the contractor three sets of architect's plans and specifications and the application for the septic field. The contract provided that the owner may order the contractor to stop work until the cause of the stop order is eliminated. The contract also provided for remedies upon the contractor's default. The contractor hired a subcontractor who filed a site plan for the septic field in an incorrect location. While the lot was being graded, the owner ordered that the swale be cut even though the general contractor told the owner that the swale would traverse the septic field. The health department then revoked the septic permits. The general contractor met with the owner to obtain a partial payment under the contract. Meanwhile, the village posted a stop work order on the property due to the revoked septic permit. Upon learning this, the owner stopped payment on the check and effectively fired the general contractor. The general contractor denied that it knew of the stop work order when it collected the check. The court affirmed the judgment for the general contractor, holding that the owner's attorney's letter to the general contractor stopping work was an attempt to terminate the contract and that the contractor's leaving the project following the letter effected a rescission of the contract, whereupon the contractor was entitled to recover in quantum meruit.

CHAPTER 7

ACCELERATION CLAIMS

§ 7.13 Acceleration Damages

Page 190, add at end of section:

In *Transpower Constructors v. Grand River Dam Authority*, 905 F.2d 1413 (10th Cir. 1990) (applying Oklahoma law), a contractor sued the owner and its construction manager. The contractor was to build a transmission line for the dam authority, which was the project owner. It was to install poles, wires, and accessories furnished by the owner. The contract made the agent responsible for certification of applications for payment and requests for time extensions. The agent was required to design, inspect, and administer the construction. Because of a delay in the owner's delivery of the poles, the plaintiff contractor was given additional time to complete the project and received additional compensation. Bad weather and unexpected subsurface rock conditions further delayed performance of the excavation portion of the work. The agent refused to recommend that the owner grant additional time for these reasons and, following the recommendation of the agent, the owner ordered the plaintiff to accelerate progress on the project. This required the hiring of additional workers and use of additional equipment at additional cost to the plaintiff. The plaintiff submitted a claim for these additional expenses to the agent, who recommended that the claim be rejected. Thereupon, the plaintiff filed this suit against the owner for breach of contract and against the agent for negligence. The court affirmed the judgment entered on the jury verdict for the plaintiff, holding that the verdict was supported by the evidence. Damages could be awarded even though the plaintiff did not meet the elements of a "total cost theory" and it was not mathematically possible to precisely apportion damages to various causes.

CHAPTER 8
DIFFERING SITE CONDITIONS CLAIMS

§ 8.3 Typical Clauses

Page 204, add at end of section:

In *Ruby-Collins, Inc. v. City of Charlotte*, 740 F. Supp. 1159 (W.D.N.C. 1990), the contractor on a city water main construction project sought an equitable adjustment of the contract due to increased costs. During construction of the water main through the city, the contractor was required to construct a trench for the pipeline, lay the pipeline, and then fill in the trench. The city retained an engineer to draft the specifications for the project. The engineer also performed soil testing. The engineering report was part of the bidding package. Based on that information, the contractor submitted the low bid and was awarded the contract. The contractor had planned to use the dirt excavated from the trench, in a typical backfilling procedure, for refilling the trench following the laying of the pipeline. This decision was made after reviewing the soil-testing data and was the basis on which the bid was made. However, when the contractor started work, it determined that, due to excessive moisture content, 20 percent of the soil excavated from the trenches could not be used for backfill. The contractor expended additional sums to bring in proper fill for the trenches. The court granted the city's motion for summary judgment, holding that the lack of a changed conditions or differing site condition clause in the construction contract effectively allocated the risk of such a problem to the contractor. The court held that the additional work to truck in the fill material due to the soil problems was covered by the original contract and therefore was not subject to equitable adjustment. The fact that the contractor compiled its bid for the project on the assumption that the soil in the relevant areas would be similar to the results of the soil testing did not amount to a mutual mistake of fact that would support contract remedies.

§ 8.15 Effect of Site Investigation and Other Disclaimer Clauses

Page 219, add to footnote 98:

A carpeting company sued the owner to recover under installation contracts in Turner Brooks v. Bowling Green State Univ., 51 Ohio Misc. 2d 1, 554 N.E.2d 956 (Ct. Cl. 1989). The bid request did not specify a completion date. The bid was accepted through a purchase order. Upon receipt of the order, the plaintiff ordered the carpet. The carpet was received five weeks later and the installation

§ 8.22 ENCOUNTERED CONDITION PHASE

commenced the next day. There were unforeseen expenses due to an abnormal floor condition that was not detectable upon the initial inspection at the time the bid was prepared. The plaintiff incurred an additional expense of $3,544 due to the floor preparation work, and the defendant claimed that it incurred an additional $7,708 for additional custodial work due to the delay in completion. The court held that the plaintiff installed the carpet within a reasonable time. The court also held that the plaintiff could recover the additional costs in floor preparation and that the defendant could not offset additional custodial expense where there was no evidence that the plaintiff failed properly to clean the areas in which it had worked.

* **§ 8.22 Proving a Claim from the Encountered Conditions Phase**

Page 233, add at end of section:

The plaintiff contractor sued the federal government for an equitable adjustment in the contract price for renovation of a walkway between buildings at a veterans medical center. In *CCM Corp. v. United States*, 20 Cl. Ct. 649 (1990), the plaintiff contractor claimed differing site conditions pursuant to the contract provisions with regard to the re-waterproofing of the enclosure. The plaintiff had made a visual inspection of the site but did not inquire about the nature of the existing waterproofing system, nor did it attend the prebid conference or the official prebid inspection. The plaintiff's bid was accepted while the plaintiff still did not know the type of waterproofing already on the concrete surface. The plaintiff used its prior experience and expected that the waterproofing system was asphalt. The bid proposal indicated that hot rubberized asphalt was to be used. The plaintiff started sawcutting the connecting areas and discovered a tar-pitch waterproofing system. The system that the plaintiff intended to use could not be applied to the surface, which was covered with tar pitch, without expensive decontamination work. The plaintiff advised the technical representative of the contracting officer and submitted a new plan for waterproofing, which was approved. Numerous weeks for re-waterproofing were lost. The plaintiff again tried to substitute another waterproofing system, which was denied, as was a change order for an extension of time and additional amounts. The court held that the plaintiff could pursue a Category One claim under the contract's provisions dealing with changes because it proved that it sustained damage from the unanticipated conditions. The fact that the delay and problems were partially the fault of the contractor was a defense to its claim for damages sought for the delays caused by the plaintiff's submission of various alternative waterproofing systems after it encountered the unanticipated conditions.

In *Servidone Construction Corp. v. United States*, 931 F.2d 860 (Fed. Cir. 1991), the contractor filed a claim for costs in excess of the amount of the bid under the terms of the Contract Disputes Act. The contractor was awarded a contract to construct an embankment, a spillway, roads, and other works on an earthen dam. The project was plagued with problems attributed to differing site conditions. The contractor filed a certified claim seeking an equitable adjustment of the contract and completed the contract work 17 months later. The contractor claimed that the Army Corps of Engineers breached an implied duty to provide adequate information on site conditions for contract performance. The contractor alleged that there were unusual soil conditions which were within the scope of the contract's site condition clause and claimed that the Corps' excessive quality assurance testing resulted in unreasonable delays. The court affirmed the Claims Court's finding for the contractor on the increased cost claim. The court found that the Claims Court did not commit error in using a modified total cost method to determine the amount of damages. However, the court found that the contractor could not recover the interest it was required to pay on money borrowed to cover the excess costs until such costs were ultimately awarded.

In *Al Johnson Construction Co. v. United States*, 20 Cl. Ct. 184 (1990), the contractor brought suit under the Contract Disputes Act to recover for differing site conditions and a constructive change in the contract. The Mississippi River Commission implemented a project to regulate the flow from the Mississippi River in certain areas. The project required construction of two water control structures which were essentially a dam and an overbank structure. Various other smaller levees, locks, and closures were also part of the project. After a major flood in 1973 damaged the dam, the Corps of Engineers determined that it was necessary to build these additional control structures. Preliminary excavation of a future channel was performed. The second phase of the project, which included construction of the control structure, involved the plaintiff. The plaintiff was to design, install, and operate a dewatering and surface water control system to assure that the area would be kept dry during construction; the plaintiff was to lower the groundwater level to at least five feet below the bottom of the excavation. The bid documents contained various test data dealing with the amount of groundwater and information from pump tests and borings. The information was designed to advise the contractor of the permeability of the soil and the quantity of water to be removed.

With respect to the claim for differing site conditions, the court said that the contractor's determination of the amount of water to be removed was not the product of a reasonable manipulation of the information in the bid materials. The contractor's projections were based on invalid assumptions which were unreasonable. The court found that the conditions encountered were not materially different from those in the contract. The court held that the contractor was entitled to recover for expenses incurred in performing additional work ordered by the Corps of Engineers which was not needed to dewater the excavation area.

Lathan Co. v. United States, 20 Cl. Ct. 122 (1990), was a contractor's suit seeking equitable adjustment. The plaintiff and NASA contracted for the removal and repair of a building roof at the Kennedy Space Center. The plaintiff was given the opportunity for a prebid inspection of the roof. Several contractors sought to cut a sample of the roof but no samples were actually taken. The bidders asked about the composition and thickness of the roof but NASA did not give

§ 8.22 ENCOUNTERED CONDITION PHASE

specific thickness measurements. While performing the contract, the plaintiff found that the existing roof was thicker than expected. The plaintiff sought additional sums due to differing site conditions, but the contracting officer denied the request. While performing the roof work, the roof started to leak, and this caused NASA to reschedule the plaintiff's operations. After it completed the contract work, the plaintiff filed a claim for equitable adjustment. Only a small portion of the claim was approved by the contracting officer. The plaintiff also claimed that the NASA changes in scheduling raised its costs. The government filed a counterclaim alleging that the plaintiff breached its contract by failing to protect the equipment in the building from leaks that occurred during the roof work. The court denied the plaintiff's motion for summary judgment, holding that there were factual questions as to whether there were differing site conditions. The contractor was required to establish that the representations made in the contract with respect to site conditions were materially different from the actual conditions. There were also factual questions as to whether the contractor or the government was responsible for the delays encountered in the project.

The problem of differing site conditions was also raised in *Husman, Inc. v. Triton Coal Co.*, 809 P.2d 796 (Wyo. 1991). The owner contracted with the plaintiff excavating company to remove overburden and topsoil from the owner's coal mine site. The plaintiff started to work and then discovered that the material to be removed was saturated with water. The plaintiff made a claim for additional amounts attributable to the unusual site conditions. The plaintiff alleged that the owner misrepresented the site conditions or intentionally concealed those conditions. The court reversed the summary judgment granted to the owner, holding that there were factual questions precluding such a judgment on the misrepresentation and concealment claims. The court, however, rejected the plaintiff's claim for recovery for breach of the implied covenant of good faith and fair dealing. The only viable claims were under misrepresentation and fraud theories.

CHAPTER 9

CHANGES IN SCOPE CLAIMS

§ 9.1 Introduction

Page 245, add to footnote 1:

Delay damages were sought by a subcontractor from the general contractor and its surety in Port Chester Elec. Constr. Corp. v. HBE Corp., 894 F.2d 47 (2d Cir. 1990) (applying New York law). The plaintiff was an electrical subcontractor involved in the renovation of a hospital. The plaintiff sought additional sums arising out of a number of design changes, errors, stop-work orders, and other actions that increased its costs of performing the job. The contract provided: "Contractor may, from time to time, modify or alter the work schedule, but in such an event, no such modification or alteration shall entitle Subcontractor to any increase in the consideration of the Subcontract." 894 F.2d at 48. The Second Circuit found this contract language to be ambiguous and reversed the trial court's dismissal of the suit. The court based its decision on the general precept that clauses seeking to exempt one from the consequences of one's own fault are to be strictly construed. The court found that the provision was not sufficiently clear and unambiguous so as to be enforceable under these facts.

In National Sand, Inc. v. Nagel Constr., Inc., 182 Mich. App. 327, 451 N.W.2d 618 (1990), a subcontractor sued the engineers and owners (sanitary drain commission) for breach of contract. The plaintiff excavated sewage pits at a sewage treatment facility. During the excavation, the plaintiff determined that there was insufficient clay on site to line the pits. An alternative site was chosen and the contracts were amended to provide for additional costs incurred due to the lack of clay at the first site. It was later discovered that the second area also had insufficient clay, and the contracts were again amended to cover additional costs. After the plaintiff completed the pits, it was paid the original contract amount, but the general contractor failed to pay additional amounts incurred due to the insufficient clay deposits. The court affirmed the lower court dismissal of the breach of contract claims against the engineers and owners, since there was no contractual relationship between the plaintiff and those parties. The lack of a contract did not, however, prevent the plaintiff from pursuing claims sounding in malpractice against the engineer. Clearly, the plaintiff did have a breach of contract action against the general contractor with whom it had contracted.

§ 9.2 Formal Change Orders

Page 247, add to footnote 6:

A contractor placed a mechanic's lien on the owner's home for work done in excess of that called for in the contract and for which the contractor was not paid.

§ 9.2 FORMAL CHANGE ORDERS

The owner moved for a summary vacation of the lien. The contractor was hired to build an addition onto the owner's home. An agreed-upon change order was included in the contract price. The written contract provided that any other modifications were to be in writing. The contractor performed extra work according to the owner's oral directions. The owner paid the contract amount and the contracter filed a mechanic's lien for $18,333 for the extra work. The court affirmed the trial court's denial of the motion for summary vacation of the lien, since there were questions of material fact as to the authorization for the extra work. Care Systems, Inc. v. Laramee, 155 A.D.2d 770, 547 N.Y.S.2d 471 (1989).

* In Thermoglaze, Inc. v. Morningside Gardens Co., 23 Conn. App. 741, 583 A.2d 1331 (1991), a contractor sued the owner of an apartment complex for breach of contract. The contractor was to furnish and install 444 windows in the complex for $88,960. The contract left the selection of the color open and provided that the owner could choose the color so as to permit the contractor to order the appropriate windows from the manufacturer. The contract was signed without a decision on the color of the windows. When the first installment was paid, the owner specified white windows. These were ordered and delivered to the site and stored at the complex until installation commenced. During that period, the owner never complained about the color of the windows. The plaintiff installed 195 windows, after which the owner's representative told the contractor that the owner's wife was unhappy with the color and wanted to change it. The owner then claimed that it ordered the other color in the first instance. The work stopped and the parties reached an impasse. A modification letter was signed for the replacement, with the contractor having the understanding that the owner would compensate the contractor for its financial loss by awarding other lucrative contracts on other properties. The court held that the modification letter was not supported by consideration and was not enforceable. The court affirmed the judgment for the contractor, finding that the problem was due to the owner's refusal to choose the color in a timely manner.

Page 247, add to footnote 7:

A contractor sought to recover sums due under a street paving contract in Bechtold Paving, Inc. v. City of Kenmare, 446 N.W.2d 19 (N.D. 1989). The city had retained an engineer to draw the plans and specifications for the project. The engineer authorized the contractor's change of materials, and approved and accepted the contractor's work, despite the fact that there were streaks and ridges in the pavement. The city was not pleased with the quality of the work and alleged that the project was not done in a workmanlike manner; it sought to recover damages in order to have the streets repaved.

The court found that the deficiencies in the paving were due to the changes in the materials, which were improper for the purpose, and because of the improper application of those materials to the road surface. However, since the engineer had approved the work, the contractor was entitled to the balance due under the contract. The court recognized that liability could be imposed on the engineer for negligence in permitting the substitution and the subsequent acceptance of the project. Thus, damages could be awarded against the engineer for the cost of repaving the street.

* *Page 247, add at end of section:*

In *Blau Mechanical Corp. v. City of New York*, 158 A.D.2d 373, 551 N.Y.S.2d 228 (1990), the plaintiff contracted to do the plumbing work for a simulated tropical rain forest in a public park. The plaintiff sought damages due to delays allegedly caused by changes in the structure of the exhibit, which were required to be approved by various municipal departments. Some change orders called for additional excavations because of subsurface conditions that differed from the city's plans and other delays were caused when members of the public demonstrated on the job site, threatening the workers and temporarily closing down the project. The court reversed the denial of the city's motion for summary judgment, holding that the delays due to the additional excavations could not support a claim for delay damages because the information given to all bidders stated that contractors could request a modification of the contract if unusual subsurface conditions were discovered that would materially affect the cost of work. Therefore, the potential delays in such a contingency were anticipated and the plaintiff was awarded a 25 percent increase in the contract price for the additional work. Similarly, the city had contractually reserved the right to make such changes in the design and structure of the exhibit. As for the delay caused by the public protesters, there was no evidence that any such delay was attributable to the city.

In *Sorrels Steel Co. v. Great Southwest Corp.*, 906 F.2d 158 (5th Cir. 1990) (applying Florida law), a subcontractor that worked on a city arts center construction project sued the general contractor. The subcontractor was to detail, fabricate, and deliver structural and other steel. The subcontract was in the form of a purchase order which provided for inspection of the fabricated steel but explicitly excluded testing. The subcontractor was not able to maintain a continuous fabrication schedule because of incomplete and inaccurate contract drawings, slow approval of shop drawings, and numerous design changes. The city then required a thorough testing program. The subcontractor agreed to the testing but sought a change order since testing had been specifically excluded from the contract and increased the production time. The subcontractor sought additional compensation for the delays and extra work. The general contractor forwarded the claims to the city and told the subcontractor that it would be advised of the progress of the claims. When the claims were not satisfactorily resolved, the subcontractor filed suit against the general contractor. The general contended that the damages sustained by the sub were due to the actions of the city and its architect and did not arise from a breach of the purchase order. The court held that the general contractor was not liable for damages sustained by the subcontractor prior to an addendum to the purchase order, which was designed to resolve the problem of the design changes and improper drawings. After the addendum, the general could be held

§ 9.3 CONSTRUCTIVE CHANGE ORDERS

liable for resulting damages. The court also held that the general contractor could be held liable for the damages attributable to the testing. This was proper despite the fact that the city never issued a change order for the testing, as the court interpreted the addendum to the purchase as an assurance by the general contractor that such a change order would be issued.

In *Cam-Ful Industries, Inc. v. Fidelity & Deposit Co.*, 922 F.2d 156 (2d Cir. 1991) (applying New York law), a subcontractor sought additional compensation for working on wooden sheeting in a trench to accommodate electrical equipment. The suit was brought against the surety, which impleaded the general contractor, under the surety's labor and materials payment bond. The general contractor was required to make payments to all claimants for labor and material used in the project, under the prime contract. The plaintiff subcontractor qualified as a subcontractor. The court held that the surety was required to make payment to the plaintiff because the general contractor-principal modified the contract when it agreed to pay the plaintiff for the extra work. The court rejected the surety's argument that it should be discharged from the obligation because the undertaking was not part of the original contract.

In *Consolidated Federal Corp. v. Cain*, 195 Ga. App. 671, 394 S.E.2d 605 (1990), the owner sued a home builder for specific performance after the builder failed to complete a home for the contract price. The contractor filed a counterclaim for the value of the labor, materials, and services performed. The parties executed a written contract for construction of the house on the plaintiff's lot. The builder failed to complete the house for the $37,152 contract price and claimed that the value of its work was almost $60,000. The difference in the cost of the home was largely attributable to change orders. The contract required that such orders be in writing and signed by both parties. However, the court held that it was proper to permit the introduction of evidence that the contractor accepted oral change orders and changed the work accordingly and that this was the cause of the increased cost. The parties, by their conduct, waived the written change order requirement. Therefore, the court affirmed the judgment on the jury verdict for the contractor.

§ 9.3 Constructive Change Orders

Page 248, add to footnote 9:

A contractor who performed site clearing operations for a shopping center sought to enforce a mechanics' lien. When the work was performed, there were no plans or specifications for the shopping center. The contractor controlled the job site with a partner of the owner on site daily. The owner repeatedly ordered different equipment and changed the work after assuring the contractor that it would pay for all the work. Nine bills were submitted: six were approved and paid, two were approved but not paid, and the last was not approved. The three were not paid under the claim that certain extra work performed by the contractor was actually within the scope of the contract. The owner claimed that those extras were actually covered under the guaranteed maximum price in the contract. The contractor then filed the mechanics' lien for the unpaid balance. The court, in Tilcon Gammino, Inc. v. Commercial Assocs., 570 A.2d 1102 (R.I. 1990), affirmed the lower court judgment for the contractor. The court held that the guaranteed maximum price applied only to work performed after the date of the document setting forth that price, and did not include work performed under the contract prior to that date. The court also found that the owner was responsible for paying for excavation work performed by a subcontractor, and that such amounts could not be charged back to the contractor.

* *Page 249, add at end of section:*

In *Allen & O'Hara, Inc. v. Barrett Wrecking, Inc.*, 898 F.2d 512 (7th Cir. 1990) (applying Wisconsin law), the plaintiff general contractor hired the defendant to perform demolition work on a portion of an office building as part of a renovation project. The contract provided for a schedule for demolition that could not be accomplished because the owner delayed three and one-half months in vacating the building. The defendant then submitted a four-month completion schedule, and the contract was amended to provide for the work to be performed over that period. The defendant encountered unanticipated conditions which resulted in further delays. The general contractor also required that the defendant use various procedures to reduce dust, noise, and vibrations. The defendant submitted a new completion schedule but, because of the delays, the plaintiff terminated the contract. Thereupon, the plaintiff sued the demolition contractor and its surety for breach of contract. The issue revolved around which party was responsible for the delays and which party should bear the cost of additional demolition. The court held that there was evidence to support a finding that the contract was modified to cover extra work claimed by the demolition contractor. The court found that the subcontract provisions requiring that change orders be in writing had been waived by the parties.

§ 9.5 Notice and Writing Requirements

Page 252, add to footnote 29:

A plaintiff contractor sued a city and the city's engineers in L. Loyer Constr. Co. v. City of Novi, 179 Mich. App. 781, 446 N.W.2d 364 (1989), after extra costs were incurred in the improvement of the city's storm drain system due to unanticipated soil and water conditions. The city refused to pay the increased costs. No agreement as to additional compensation was ever reached with the engineer who recognized the site problems and the additional work. The contract provided:

> 5.22 CHANGES IN THE WORK
>
> The Owner shall have the right, without invalidating the Contract, to require changes in, additions to, or deductions from the work. . . .

§ 9.5 NOTICE AND WRITING REQUIREMENTS

In giving instructions, the Engineer shall have authority to make minor changes, not involving extra cost and not inconsistent with the purposes of the work, but otherwise, except in emergency endangering life or property, no extra work shall be done nor change be made unless in pursuance of a written order from the Owner, signed or countersigned by the Engineer, or a written order from the Engineer stating that the Owner has authorized the extra work or change, and no claim for payment for changes in the work shall be valid except for changes so ordered. Adjustments, if any, in the amount to be paid the Contractor by reason of a change in the work shall be determined by one or more of the following methods:

    a. For changes in quantities or items of work covered by unit prices in the Proposal and Agreement, adjustments shall be made by using the actual numbers of units of work done. . . .

    b. Failing a satisfactory agreement . . . the work shall be paid for on the basis of the actual cost of labor, heavy powered equipment rental, and materials, plus 15 percent which 15 percent shall include supervision, use of hand tools, appliances, small powered equipment, salvageable material, overhead, office and general expense and all other expenses of Contractor, and profit.

                  5.23   CLAIMS FOR EXTRA COST

If the Contractor claims that any instructions by drawings or otherwise involve extra cost under this Contract, he shall give the Engineer written notice thereof within a reasonable time after the receipt of such instructions, and in any event before proceeding to execute the work, except in emergencies endangering life or property, and the procedure shall then be as provided for changes in the work. No such claim shall be valid unless so made.

446 N.W.2d at 366-67.

The court reversed the lower court's summary judgment for the defendants on negligence and misrepresentation claims. The court held that the contactor's extra cost claim was limited to the unit prices for the excavation. No compensation was awarded for the additional work in depressurizing aquifers or for the construction of berms. The court also held that the engineer could not be held liable on the contractor's breach of contract claim, because the engineer was not a party to that contract.

Page 252, add to footnote 30:

The Soil Conservation Service (SCS) was asked to assist a local reservoir control district following severe flooding. The SCS prepared a plan and contract for restoration of the damaged areas. Surveys were made and stakes placed for the limits of the area to be excavated. An SCS engineer prepared cubic yard estimates of the amount of materials to be excavated for sediment removal and dike restoration. The bid specifications provided for payment on a linear foot basis. No determination was made at the time of contracting as to which portions would only have to be repaired and which sections would have to be replaced. Several modifications were agreed upon during the course of the work. Two certified claims were approved by the contracting officer and one was denied. It was claimed that additional amounts should have been paid for extra dike restoration and sediment removal.

The court, in Spirit Leveling Contractors v. United States, 19 Cl. Ct. 84 (1989), entered judgment for the SCS. The court held that certain items for which recovery was sought in the suit had not been properly presented to the contracting officer. This was a jurisdictional defect. The court rejected the contention that extra work was required due to site and quantity variations which the contractor claimed were caused by weather conditions. The court also dismissed all claims contained within the scope of a release given by the contractor to the government which were not set forth in a certified claim letter and which were submitted after final payment was made under the contract.

Page 252, add to footnote 32:

In Cable Belt Conveyors, Inc. v. Alumina Partners, 717 F. Supp. 1021 (S.D.N.Y. 1989), there was a contract to build and install a conveyor system running from the owner's bauxite mine to a storage dome. The subcontractor was to perform the installation under the terms of its subcontract. All contracts provided for the arbitration of disputes. The subcontractor subsequently sought arbitration for claims totalling $33 million. The subcontractor sought the cost of changes and delays caused by the general contractor. The general contractor sought arbitration from the owner for any portions of the subcontractor's claims for which it could be held liable. The general contractor and subcontractor agreed to cooperate in pressing their claims against the owner and to divide any amounts recovered. The court held that the agreement between the owner and the general contractor did not bar the general contractor from passing through the claims of the subcontractor. The court also consolidated the arbitration under the contract and the separate arbitration under the subcontract in order to avoid inconsistent results.

Page 253, add to footnote 35:

Westates Constr. Co. v. City of Cheyenne, 775 P.2d 502 (Wyo. 1989) involved a water diversion project. The project was subject to control by the United States Forest Service with regard to pollution and soil erosion aspects. As part of the work, the subcontractors washed gravel and drained muddy water from the project into holding ponds. The ponds failed and muddy discharge flowed into a creek. The Forest Service was concerned that the project would affect spring runoff and suspended work until it determined what protective measures were required. It also required the contractor to perform extra work, which delayed the project and resulted in increased costs. The contractor sought to recover these costs from the city. The court affirmed summary judgment for the city on the issue, holding that the contractor waived its right to seek additional compensation when it failed to submit a substantiated claim as per the contract.

* *Page 254, add after carryover paragraph:*

The plaintiff contractor was required to perform additional work to replace exhaust fans in a government building and sued for a price adjustment, in *Robert*

§ 9.5 NOTICE AND WRITING REQUIREMENTS

Irsay Co. v. United States Postal Service, 21 Cl. Ct. 502 (1990). The plaintiff contracted with the Postal Service to replace exhaust fans. During work, the plaintiff advised the contract manager of unforeseen lead lining some of the ductwork. The plaintiff told the manager:

> This situation, not contemplated at the time of contracting, has and will continue to have an adverse effect on our performance of this part of our contract work.
>
> If we are forced, because of this lead coating and lead lining situation, to resort to dismantling these fans using wrenches, chisels and sawcuttings, we anticipated [sic] that the cost of performing this work will be severely impacted as well as extending the time required to accomplish the work.
>
> At the present time, we are trying to determine, what must be done with respect to this problem of getting the work done while protecting the health and welfare of our (and your) personnel. With this in mind, and to insure that our work crews are not subjected to health hazards as a result of airborne lead and other metal fumes we have brought in Environmental Consultants (see attached letter) to conduct tests on this work environment.
>
> * * *
>
> Accordingly, by this letter, you are advised, that this unknown condition is having an adverse effect on our cost of performing this work, and the time for its performance and when these additional costs are known, we will submit our proposal for these additional costs.

Various follow-up letters were sent to the manager detailing the additional expenses incurred. The court held that these letters did not comply with the requirements of the Contract Disputes Act, as they were never sent to the contract officer. The court held that the jurisdictional prerequisites cannot be waived and were not met in this case.

In *Al Johnson Construction Co. v. United States*, 19 Cl. Ct. 732 (1990), the government sought to dismiss the suit on the basis of the Contract Disputes Act. The plaintiff was a joint venture which contracted with the Corps of Engineers to construct a dam and various channels. Disagreements arose during construction with respect to the responsibility for the design of concrete mixes. The plaintiff claimed that it incurred additional costs because of the government's poor handling of the mix design. The plaintiff submitted a certified claim for the costs, which was signed by the project manager, to the contracting officer. The question was whether the claim was properly certified under the terms of the Act. The court dismissed the suit, finding that the project manager was not a senior company official as required by the statute and the applicable regulation (32 C.F.R. § 7-103.12). In the absence of a proper certification, there was no viable claim.

Lakeview Construction Co. v. United States, 21 Cl. Ct. 269 (1990), was a suit by a contractor seeking to recover impact and delay costs occasioned by change orders. During performance of the contract, the Navy sought certain changes. The contractor reserved its right to claim impact and delay costs for each of the 22 modifications made pursuant to change orders. The question was whether the various paperwork submissions complied with the requirements of the Contract Disputes Act (41 U.S.C. § 601 *et seq.*). The contractor's paperwork consisted of various letters with attachments signed by officers of the contractor and its counsel. Some of the letters did not specifically demand a decision on the claim by the contracting officer. The court set out the language of various letters, some of which did request a final decision by the contracting officer. The court granted the government's motion to dismiss the complaint due to the failure of the contractor to submit the claims to the contracting officer. Instead, the contractor had presented the claim to the resident officer in charge of construction. This was not in compliance with the notice requirements of the statute. This case points out the need for strict compliance with the statutory requirements.

In *American Pacific Roofing Co. v. United States*, 21 Cl. Ct. 265 (1990), the plaintiff contracted with the Navy to install a new waterproof roof. After the roof was installed, it continued to leak and the plaintiff was directed by the Navy to repair the leak. The plaintiff submitted a claim for repair costs requesting a final decision from the contracting officer; however, the claim was sent to the resident officer in charge of construction, as per the contract. The claims were forwarded and denied by the contracting officer on the basis that the plaintiff failed to provide a leakproof roof as required under the contract. The plaintiff put forward another claim alleging that payments were withheld, which was also denied by the contracting officer. The plaintiff then filed suit, whereupon the government counterclaimed for reprocurement costs and water damage. The government claimed that the court lacked jurisdiction under the Contract Disputes Act because the claim had not been properly submitted to the contracting officer. The court denied the government's motion to dismiss for lack of jurisdiction, finding that the contractor complied with the terms of the Act by putting the claim in writing and submitting it to the contracting officer, even though it was transmitted through the resident officer first per the terms of the contract.

Woodhaven Homes, Inc. v. Kennedy Sheet Metal Co., 304 Ark. 415, 803 S.W.2d 508 (1991), was a suit by a subcontractor seeking payment for extra work performed in connection with the construction of a restaurant. The subcontractor was to perform the plumbing, heating, air conditioning, and ventilation work and to supply various materials. The subcontract provided that the general contractor could direct the subcontractor to change or modify its work, whereupon the sub would submit in writing its claims or adjustments in the contract price. The subcontractor was a licensed heating, air conditioning, and plumbing contractor but was not licensed as an electrical contractor. The general made various changes throughout the construction period, but did not draw up change orders. There was also a verbal agreement that the subcontractor would perform certain electrical work for a specified sum and a change order was executed for extra electrical work. At the end of the work, the subcontractor submitted bills to the general contractor for the electrical and plumbing extra work, which the general contractor refused to pay. The court affirmed the judgment for the subcontractor, permitting it to recover in quantum meruit. After the work was completed, a new contracting licensing statute (Ark. Code Ann. § 17-22-103) was enacted, providing that an unlicensed contractor could not maintain a suit in quantum meruit. The court held that the statute did not apply to work already performed.

§ 9.9 EXTRA AND ADDITIONAL WORK

In *A.G. Lichtenstein, P.E. v. Goldin*, 166 A.D.2d 328, 560 N.Y.S.2d 780 (1990), the plaintiff engineering consulting firm was under contract with the city to design improvements to a viaduct. The plaintiff sued the city and the city comptroller to recover for additional labor and materials costs incurred in connection with design changes requested by the Landmarks Commission after the agreed-upon work had been performed. The court affirmed the dismissal of the claim as time-barred. Under the plaintiff's contract with the city, it was required to commence any suit within six months of accrual of the claim. The court held that the comptroller's rejection of the plaintiff's claim started the running of the period. That rejection was final and binding when the comptroller never granted the plaintiff's application for reconsideration. Any further action taken by the comptroller was viewed as merely being for the purpose of potential settlement negotiations and to support the prior denial.

§ 9.6 Pricing a Change Claim

Page 255, add after third sentence of first paragraph:

In *Markway Construction Co. v. Kirchenbauer*, 769 S.W.2d 836 (Mo. Ct. App. 1989), a remodeling contractor sued to recover the balance due under its remodeling contract. The contractor also sought to recover on several unsigned change orders. The owners claimed that the contractor failed to perform in a workmanlike manner, and they also sought delay damages. The contract provided for payment on a cost-plus basis, and stated that "the Owner may make Changes in the Work as provided in the Contract Documents. The Contractor shall be reimbursed for Changes in the Work on the basis of Cost of Work as defined in [the contract]." *Id.* at 838. The court held the owners liable for the cost of the change orders, since the contract did not require that the orders be in writing or signed by the owner. The owners accepted the work from the change orders without complaining that it was not authorized. The owners, however, were entitled to recover the costs incurred due to the contractor's delay in performance.

§ 9.9 – Extra and Additional Work

Page 260, add to footnote 64:

In City of Jacksonville v. W.R. Fairchild Constr. Co., Ltd., 547 So. 2d 1010 (Fla. Dist. Ct. App. 1989), the contractor was to excavate and lay ductile iron pipe. The contract required that the contractor drive steel sheeting on both sides of the excavation for safety reasons. The question arose as to whether the contractor was to be paid for the driving and pulling of the steel sheeting in addition to the installation price of the pipe (the steel sheeting was all pulled from the excavation after it was no longer needed). The court held that the contract did not provide that the driving and pulling of the sheeting was a separate item of work for which the contractor would be compensated. The contract provided that only sheeting left in place was separately compensable, and in fact, all sheeting had been removed per the drawings and contract documents.

In Acquisition Corp. v. American Cast Iron Co., 543 So. 2d 878 (Fla. Dist. Ct. App. 1989), a supplier filed a lien against the project and a subcontractor in connection with a sewer and drainage project. The subcontractor also brought claims against the general contractor. The subcontractor was required to obtain and provide certain materials listed on the plans. Additional quantities to those listed were to form an additional contract, but no such extra work or materials were to be compensable unless authorized in writing by the general contractor. The plaintiff supplier delivered the required iron pipe to a subcontractor who subsequently walked off the job shortly before completion. The supplier was not paid and filed a claim of lien. The subcontractor also sued the contractor to recover for extra work performed due to improper preparation by the general contractor. The court held that the subcontractor was entitled to recover for the extra work caused by this problem and an additional requirement imposed by the health department. The subcontractor recovered the amounts due under the contract and for extra materials supplied due to errors in the plans prepared by the owner. This, in turn, provided payment to the supplier. The court refused to permit the subcontractor to recover for design changes which occurred prior to the beginning of the construction work where there was no compliance with the contract requirement that extra work be authorized in writing. Such a writing was not required where the extra work was due to health department dictates or errors in the plans.

Page 260, add to footnote 65:

A property owner wanted to have a pond excavated on his property. He contracted with the plaintiff to perform the excavation work and to grade a hillside. The contractor was to be paid $1.50 per cubic yard of earth moved and $1,000 to $1,500 for the grading work. The owner had the Soil Conservation Service sketch a plan and stake the pond outline. The area was 100 feet square and was to be excavated to a depth of 12 feet. The contractor estimated that this would require the moving of 2,000 cubic yards of earth, and the cost for that would be approximately $3,000. The owner repeatedly moved the stakes and ended up with a pond 100 feet by 200 feet and 12 feet deep. This change doubled the excavation cost. The owner stated that he had agreed to pay $3,000, which represented the original amount to be excavated and not the additional cost incurred by the changes in the stakes. In Gorbett v. Claycamp, 553 N.E.2d 475 (Ind. 1990), the court held that the contractor was entitled to the $1.50 per cubic yard rate multiplied by the 4,000 cubic yards actually removed. The court held that the contractor was not bound by the estimate where the owner changed the dimensions of the area to be excavated.

Page 260, add to footnote 66:

In a case involving two disputes between a mechanical contractor and a subcontractor which was to design, furnish, and install regulating devices, questions

§ 9.9 EXTRA AND ADDITIONAL WORK

of contract interpretation arose as to whether the subcontractor was required to furnish and install certain controls. The subcontractor contended that such work constituted extras for which additional compensation was required. The contract was a printed form, but, under the provisions dealing with duties of the subcontractor, there was a handwritten clause. The court determined that the handwritten provision was not intended to exclude certain controls from the scope of the work to be performed by the subcontractor. This required the subcontractor to provide the controls without additional compensation. However, with respect to another dispute, the approval of extra-work invoices by the mechanical contractor's foreman supported the recovery of such amounts, despite the claim that the invoices were not the final determination of whether such a claim for additional compensation was valid. Apex Control Sys., Inc. v. Alaska Mechanical, Inc., 776 P.2d 310 (Alaska 1989).

* *Page 260, add at end of section:*

In *Garcia v. Kastner Farms, Inc.*, 789 S.W.2d 656 (Tex. Ct. App. 1990), the owner contracted with a general contractor for construction of an agricultural irrigation reservoir. The contract called for excavation of a large volume of clay to seal the reservoir walls. This work was subcontracted to the plaintiff. The contract set payment at $1.40 per cubic yard of naturally compacted material hauled to the job site. The plaintiff began to haul clay to the construction site. However, the local police ticketed the trucks for overloading. The truck drivers refused to continue hauling the clay after they received citations, since at the $1.40 price to carry smaller loads was unprofitable. The owner complained that more clay was required. The truckers were unreliable and various shortages occurred, resulting in project disruptions. The owner offered to pay $1.75 per cubic yard if more trucks were brought in. After this price change, the hauling proceeded without incident. The appellate court held that the owner had not ratified the increase in the rate per cubic yard where the change was the product of duress. There was testimony that the owner would never have offered such a high rate except for the fact that no other truckers were available after the project started. The owner only acquiesced after the truckers decided that they would no longer perform at the agreed-upon rate provided for in the written contract which governed the rate. However, the contractor was entitled to be paid on the basis of quantum meruit.

A contractor brought suit to recover additional amounts in addition to the contract price for extra work performed. The contractor was hired to modify the heating system for a three-flat apartment building. The contractor was to cut down the existing boiler so that it would heat only the first floor, install boilers to heat the other two floors, and connect each boiler to separate gas meters. The contractor's alleged extra work included the installation of steam pipelines running from the newly installed boilers to the radiators in the upper apartments. The appellate court in *Duncan v. Cannon*, 204 Ill. App. 3d 160, 561 N.E.2d 1147 (1990), held that the plaintiff failed to carry its burden of proving any entitlement to additional compensation. There was insufficient proof that the pipelines were outside the scope of the original contract.

CHANGES IN SCOPE CLAIMS

In *Rembrant, Inc. ex rel. Wright Construction Co. v. United States*, 919 F.2d 1569 (Fed. Cir. 1990) (applying Contract Disputes Act), the contractor applied a third coat of paint to the buildings in question, on the order of the government, in accordance with the government's interpretation of the contract. This suit was brought to determine whether the contract required application of the third coat or whether there was an implied exception for previously painted surfaces which required only two coats. The painting schedule in the contract had 38 separate items for different surfaces, 26 of which specified "None" for the "3rd Coat." The surfaces listed in the painting schedule included wood, ferrous, concrete, metal, and plaster. The contract provided:

> PAINTING SCHEDULE: THE PAINTING SCHEDULE prescribes the surfaces to be painted, required preparation and the number of coats of paint. The schedule specifie[s] the type of paint for three coats.

The subcontractor viewed the additional painting, the third coat, as constituting extra work not required under the contract and submitted claims to the general contractor for the extra cost. The court held that the contract unambiguously called for three coats and did not exempt previously painted surfaces so as to require only two coats for such surfaces. Therefore, the subcontractor was not entitled to additional compensation.

In *Owners Realty Management Construction Corp. v. Board of Education West Islip*, 160 A.D.2d 921, 554 N.Y.S.2d 648 (1990), the plaintiff asbestos removal contractor sought to recover for additional work, outside the scope of the contract, performed at the request of the defendant. This extra work was requested by the board's architect and assistant supervisors in areas that were not addressed in the contract. The court affirmed the denial of the board's motion for summary judgment, holding that there were factual questions as to whether the work in question was outside of the contract.

In *Western Empire Constructors, Inc. v. United States*, 20 Cl. Ct. 668 (1990), the plaintiff was the contractor on a project to remodel a Veterans Administration hospital. It sought an equitable adjustment of the contract price. Prior to bidding, the plaintiff visited the hospital and had an opportunity to inspect the position of existing switch boxes. The contract included drawings and specifications for the light switches and boxes. The project called for the conversion of the entire electrical system to a higher voltage. Therefore, all existing wall switches were to be removed and relocated pursuant to the contract drawings. The drawings showed where the new boxes were to be located but not where the existing boxes were. The Veterans Administration later clarified the height of the boxes and stated that they could remain at the existing height in only a few areas. A total of 1,262 boxes were to be lowered. There was a dispute as to whether the contract called for the lowering of the boxes. The plaintiff submitted a claim for the cost of lowering the switches and boxes. The contracting officer denied the claim and found that the contract drawings properly informed the plaintiff of the relocation requirement. The court denied the claim, holding that an equitable adjustment

§ 9.10 AUTHORIZATION

was not warranted, as it found that the contract clearly required the lowering of the boxes to accommodate the handicapped.

§ 9.10 – Authorization

Page 260, add to footnote 67:

The case of Design & Production, Inc. v. United States, 18 Cl. Ct. 168 (1989), was brought under the Contract Disputes Act (41 U.S.C. § 601 *et seq.*), with respect to a public construction contract. The plaintiff contractor had filed a certified claim and an amended certified claim with the appropriate contracting officer, seeking money for additional work. The contract officer denied that claim on the basis that the work was within the scope of the original contract. In response to the contractor's claim, the government filed a counterclaim contending that the contractor failed to supply all equipment and materials called for in the contract.

The court held that the contractor was entitled to an equitable adjustment based on additional work which was authorized by an appropriate government official. The government was unsuccessful on its counterclaim. The court was required to engage in contract interpretation and admitted the use of parol evidence in instances in which the contract was ambiguous. The court found that the contract did not require the construction of certain walls that had been ordered at the discretion of the government. Such work was extra and supported additional compensation. However, certain claims of the contractor relating to extra work were rejected. The court found that the additional costs, incurred because the contractor was required to perform more work than it had anticipated to fulfill the contract, were not subject to equitable adjustment.

Gymco Constr. Co. v. Architectural Glass & Windows, Inc., 884 F.2d 1362 (11th Cir. 1989) (applying Georgia law), arose out of the construction of an exercise salon. A mirrored glass facade was to be installed. The contract specified a completion date following the receipt of approved shop drawings and the delivery of materials. A template was to be provided for the location of holes to be drilled for the anchoring of a sign. The template was to be created by another subcontractor, the sign company. The template arrived five weeks late, which did not permit the installation of the glass by the contract completion date.

A salesperson made an oral agreement to switch from the glass to stainless steel. Upon delivery of the steel, the contractor refused to install the steel, which was more expensive than the glass and was outside the company's expertise. Any such agreement between the general contractor and the subcontractor's salesperson was clearly outside the scope of the salesperson's authority. The general contractor contracted with another subcontractor to supply and install the steel panels. The price of the steel and its installation was $50,414 higher than for the glass. The general contractor then sued the glass subcontractor for the difference, in a breach of contract action.

The subcontractor claimed that the general contractor had interfered with its performance of the contract. The court held that the glass subcontractor had not breached the contract by failing to install the glass in a timely fashion, since it did not receive the template on schedule. This delay constituted a breach of contract by the general contractor. However, the fact that the subcontractor continued to perform other work on the project constituted a waiver of the subcontractor's right to terminate the contract for the general contractor's breach. The subcontractor nevertheless retained its claims for damages caused by the delay in furnishing the template. Therefore, the court reversed the trial court's award of damages to the general contractor, and the case was remanded.

§ 9.11 – Reason for the Extra Work

Page 263, add to footnote 78:

In City of San Antonio v. Forgy, 769 S.W.2d 293 (Tex. Ct. App. 1989), a contractor sought to recover costs incurred in performing extra work. The contract involved the drilling and completion of a water well pursuant to plans and specifications. The plaintiff subcontracted the actual work to another. As the well was nearing completion, the metal casing ruptured. It was not possible to save the first well, and another well was drilled and completed. The plaintiff contractor sought to recover the additional costs attributable to the second well. The city repeatedly denied the contractor's request for a change order that would include the extra expense. The court held that the contractor was not entitled to the additional expense, as the city did not breach any duty of good faith owed to the contractor, even though its engineers recalculated the specifications of the casing and determined that the thickness of the casing as specified might be insufficient. The court also rejected the contractor's negligence claim, as the jury found that the contractor was 65% negligent and the city only 35% negligent.

A plaintiff contractor sought to recover additional expenses incurred in a subway construction project, arising out of unforeseen subsurface conditions, in Thomas Crimmins Contracting Co. v. City of N.Y., 74 N.Y.2d 166, 542 N.E.2d 1097, 544 N.Y.S.2d 580 (1989). The contract provided for arbitration by the project engineer of such claims. The added expenses arose out of unanticipated underground water, rock formations, and soil subsidence. The contractor filed claims for additional costs due to these circumstances and sought money for extra work required to be performed which was outside the contract. Many of the claims were accepted. The plaintiff sued to recover amounts attributable to the rejected claims. The court held that the arbitration provision in the contract was ineffective to make the engineer's decision the final authority. The court found that the claim was not subject to dismissal on the basis of the contract provision which allegedly precluded judicial review of the claim.

In Shacocass, Inc. v. Arrington Constr. Co., 116 Idaho App. 460, 776 P.2d 469 (1989), the plaintiff sought to recover for additional work performed. The plaintiff subcontractor contracted to work on the construction of sets of storage bins on a site used by the Department of Energy for the storage of nuclear waste.

§ 9.11   REASON FOR EXTRA WORK

The defendant was the cement contractor with the responsibility for the cement work on the bins. It entered into a fixed-price subcontract with the plaintiff to install steel reinforcing bars that would be embedded in the cement. During the construction process, the plaintiff became involved in a dispute with other contractors over the concrete cover specifications. The subcontractor also questioned the way in which the contractor was performing the cement work, because its method required the plaintiff to redo its rebar work. The defendant rejected the plaintiff's claim for additional compensation. The court affirmed the trial court's summary judgment for the defendant, holding that the plaintiff was not entitled to additional compensation even though it was required to redo its work to meet the contract tolerances. The contract tolerances were not unconscionable. The fact that the plaintiff incurred costs in excess of its expectations was insufficient to support a tort claim and, since there was a written contract covering the work, there could be no recovery in quantum meruit.

In McDevitt & Street Co. v. Marriott Corp., 713 F. Supp. 906 (E.D. Va. 1989), the contractor sued the owner to recover amounts due under the construction contract. In response to that claim, the owner brought a counterclaim for delay damages. The contract provided that the contractor would complete construction so that the owner could start business within 330 days of the specified commencement date. Due to a variety of factors, the hotel was completed 19 weeks late, thereby triggering a penalty clause. The case revolved around the causes of the delay and whether additional work was requested by the owner.

The court held that the contractor could not recover additional amounts for extra work related to soil conditions, because the contract specifically placed such a risk on the contractor. Any delay due to adverse weather conditions was also to be borne by the contractor. The owner, however, had constructively agreed to pay for additional soil work, but its denial of time extensions was proper and could not support a claim for other additional costs incurred by the contractor in attempting to meet the contractual time schedule. The court evaluated a variety of claims relating to the delay and additional costs, but the contractor was not excused from the delay and was not entitled to increased compensation. The court held that delay damages are to be determined with regard to the rental value of a completed building or a reasonable return on investment. Under this measure of damages, the owner was awarded over $310,000 for the 132-day delay.

In another case, Edwards v. United States, 19 Cl. Ct. 663 (1990), a contractor sought an equitable adjustment of the contract price for construction of a post office. The contract called for the contractor to build and lease the post office to the government. The Postal Service real estate specialist relied on the representations of the local postmistress that land owned by her family was suitable and appropriately zoned. In reality, the title to the land was tied up in probate, and it was not properly zoned for the proposed use. The bid invitation warned bidders that they were responsible for determining that the property was properly zoned. Further, the bidder was told that it must own or control the property. The postmistress told the successful bidders that her property was suitable. The contractors had already built one post office, but were not contractors by profession. The plaintiff

CHANGES IN SCOPE CLAIMS

contractors failed to check the title to the land in question and its zoning. The court held that they were not entitled to an equitable adjustment because zoning requirements necessitated the purchase of additional land in order to build the post office. Any such increased cost or damage sustained by the plaintiffs was due to their own negligence.

* In Prairie Land Constr., Inc. v. Village of Modesto, 213 Ill. App. 3d 364, 571 N.E.2d 1210 (1991), a contractor sued the village for work it performed in locating and checking for a suspected leak. The contractor had constructed a water line. The project was substantially completed when the village told the contractor that there appeared to be a leak in one of the pipes it had installed. The contractor checked the area in question and found no leak. The contractor billed the village $1,158.50 for the cost of the additional work. At the time of the work, the job was still under warranty. The contract provided:

> Should it be considered necessary or advisable by the Local Public Agency at any time before final acceptance of the entire work to make an examination of work already completed by uncovering the same, the Contractor shall on request promptly furnish all necessary facilities, labor, and material. If such work is found to be defective in any important or essential respect, due to fault of the Contractor or his subcontractors the Contractor shall defray all the expenses of such examination and of satisfactory reconstruction. If, however, such work is found to meet the requirements of the Contract, the actual cost of labor and material necessarily involved in the examination and replacement, plus 15 percent of such costs to cover superintendence, general expenses and profit, shall be allowed, if completion of the work of the entire Contract has been delayed thereby, [and the Contractor shall] be granted a suitable extension of time on account of the additional work involved.

The court held that it was the plaintiff's duty to detect any leak. The court found that the quoted contract language was ambiguous and the plaintiff contractor failed to meet its burden of proof on establishing its right to recover thereunder.

* In Maitland Bros. Co. v. United States, 20 Cl. Ct. 53 (1990), a contractor sued the government to recover the cost of reexcavating an underwater trench and reinstalling pipeline. The plaintiff had a contract with the Navy to dredge an underwater trench in which the plaintiff was to install submarine oil and petroleum pipelines. The contract provided for the route and depth of the trenches and pipelines. Drawings illustrated the contract requirements. The contract also provided for the allocation of responsibility and procedures for the work. The contract stated:

> The support and stress control of the pipe during the laying, pulling or connecting process shall be in accordance with generally accepted engineering practice specified in API RP 1111 and shall be the entire responsibility of the Contractor.

The contract further provided for the method of installation:

> Backfill in the pipeline trench shall be applied immediately after approved hydrostatic testing of the submarine pipeline is completed.

§ 9.12 VALUE OF EXTRA WORK

An architectual engineering firm hired by the Navy designed the pipelines for the conditions in the river. Problems with contract performance arose when the Corps of Engineers determined that the trench was not at the correct depth. The plaintiff contended that it dredged the trench to the proper depth but maintenance dredging in the shipping canal caused the discrepancy. Maintenance dredging by the Corps of Engineers also disrupted the plaintiff's work by causing sediment to accumulate in the trench. The Navy then agreed to keep a clearance between the dredging and the plaintiff's trench. This dredging disrupted the Corps of Engineer's earlier readings. After the plaintiff decided to reexcavate the trench, a subsequent test by the Corps again showed the trench to be too shallow, although a diver checked the depth and found that the pipeline was laid at a proper depth. The Armed Services Board of Contract Appeals rejected the plaintiff's claim for the reexcavation expense. The court, however, held that the evidence supported a finding that the pipeline was initially laid at an improper depth which required the reexcavation.

§ 9.12 – Value of the Extra Work

Page 263, add to footnote 79:

A city terminated a general contractor's work after it was determined that the contractor had not obtained the necessary wetlands excavation permit. The contractor sued to recover amounts for additional work and for delay damages in Earthbank, Inc. v. City of N.Y., 145 Misc. 2d 937, 549 N.Y.S.2d 314 (1989). The contract called for the contractor to be paid a unit price for excavation of the site, leveling, and refilling of the site with fill that did not contain the weed *phragmites communis*. The contractor excavated thousands of cubic yards of soil in excess of the engineer's estimate before work was stopped due to the lack of a wetlands permit. The city ordered the contractor to restore the area. A change order was issued and the contractor was paid an additional sum for that work. The contractor sought to recover sums, in addition to the unit price, for other extra work, contending that the extra work was different in quality from that provided for in the contract. The court denied the city's motion to dismiss the contractor's claim for damages above the unit price. The court further held that the contract clause purporting to eliminate delay damages did not bar the contractor's delay damage claim, which was based on the city's alleged misrepresentation that it had obtained the necessary wetlands excavation permit.

In Ken's Constr. Co. v. Liles, 560 So. 2d 103 (La. Ct. App. 1990), a contractor submitted a bid to do renovation work at the owner's home. After a bid in the amount of $55,300 was submitted, the contractor met with the owner to discuss proceeding on a cost-plus agreement. They reached an oral agreement to do the work as set out in architectural plans provided by the owners on a cost plus 10% overhead and 10% profit basis, subject to a cap in the amount of the original bid. On the first day of work, the owner told the contractor that he wished additional work and said he wanted to go cost-plus. The question was presented as to whether the job was then transformed into a straight cost-plus basis for all the work, or whether it was subject to the original cap and cost-plus for the extra work. No writing was ever made. The court held that the work was done on a straight cost-plus basis without the cap. The contractor was also entitled to interest on the amount due under the contract.

§ 9.13 Methods of Calculating Costs

Page 264, add to footnote 83:

A property owner wanted to have a pond excavated on his property. He contracted with the plaintiff to perform the excavation work and to grade a hillside. The contractor was to be paid $1.50 per cubic yard of earth moved and $1,000 to $1,500 for the grading work. The owner had the Soil Conservation Service sketch a plan and stake the pond outline. The area was 100 feet square and was to be excavated to a depth of 12 feet. The contractor estimated that this would require the moving of 2,000 cubic yards of earth, and the cost for that would be approximately $3,000. The owner repeatedly moved the stakes and ended up with a pond 100 feet by 200 feet and 12 feet deep. This change doubled the excavation cost. The owner stated that he had agreed to pay $3,000, which represented the original amount to be excavated and not the additional cost incurred by the changes in the stakes. The court, in Gorbett v. Claycamp, 553 N.E.2d 475 (Ind. 1990), held that the contractor was entitled to the $1.50 per cubic yard rate multiplied by the 4,000 cubic yards actually removed. The court held that the contractor was not bound by the estimate where the owner changed the dimensions of the area to be excavated.

In Markway Constr. Co. v. Kirchenbauer, 769 S.W.2d 836 (Mo. Ct. App. 1989), a remodeling contractor sued to recover the balance due under the remodeling contract. The contractor also sought to recover on several unsigned change orders. The owners claimed that the contractor failed to perform in a workmanlike manner, and they also sought delay damages. The contract provided for payment on a cost-plus basis, and stated that: "the Owner may make Changes in the Work as provided in the Contract Documents. The Contractor shall be reimbursed for Changes in the Work on the basis of Cost of Work as defined in [the contract]." *Id.* at 838. The court held the owners liable for the cost of the change orders, since the contract did not require that the orders be in writing or be signed by the owner. The owners accepted the work from the change orders without complaining that it was not authorized. The owners, however, were entitled to recover the costs incurred due to the contractor's delay in performance.

Page 264, add to footnote 86:

In Weaver-Bailey Contractors, Inc. v. United States, 19 Cl. Ct. 474 (1990), the contractor was awarded a contract for the construction of beaches, breakwaters, boat ramps, parking areas, and other recreational improvements on Arcadia Lake

§ 9.13 CALCULATING COSTS

in Edmond, Oklahoma. The work largely entailed grading, cutting slopes, finishing, and the construction of a breakwater (using loose rock) on the slopes bordering the lake. The Army Corps of Engineers' estimate was that it would be necessary to excavate 132,000 cubic yards of materials of varying composition. During the course of the work, it was determined that the estimate was off by ±1% and that 186,695 cubic yards would have to be excavated. The original rate of pay was $3.42 per cubic yard and a contract modification was issued to pay the contractor $3.29 per cubic yard for the additional 54,695 cubic yards. The original work schedule was disrupted by the additional excavation work and not all of the work could be completed before winter. Winter weather caused damage by erosion to some of the work, which thus required repair in the spring. The court held that the delay in the completion of the work was excusable under the facts of the case, since it arose from unforeseeable causes beyond the control of the contractor. Because of the differing site conditions, the contractor was entitled to an equitable adjustment of the contract price, based on a 10% profit rate, as it was established that without the additional excavation work the project would have been completed early.

* *Page 267, add at end of section:*

The general contractor in *HOH Co. v. Travelers Indemnity Co.*, 903 F.2d 8 (D.C. Cir. 1990), sued the subcontractor's surety over several contract disputes involving construction of a new terminal at the city airport. A performance bond was issued for the subcontractor covering "all the undertakings, covenants, terms, conditions and agreements of any and all duly authorized modifications of the Subcontract." After various disputes, the general contractor and the city executed a formal modification of the contract which resolved all past claims. The general contractor was entitled to a 14 percent markup on the cost of each additional change. The general contractor then executed a formal modification of the subcontract which resolved claims and extended the benefits of the general contractor modification to the subcontractor. There was a dispute as to whether the subcontractor's allocable share of any adjustment in the general's contract with the city would be determined by the general. Another dispute arose with respect to increased costs due to change orders. The general contractor refused to submit the claim to the city. The subcontractor alleged that the failure to present its claim constituted a breach of the subcontract. A pass-through agreement was executed and the general contractor agreed to prosecute the subcontractor's claims in return for reimbursement of all legal fees incurred in the prosecution. The general contractor filed suit on behalf of itself and the subcontractor against the city seeking the outstanding claims plus the general's 14 percent markup on the claims. The general contractor's claims on its own behalf were dismissed and ultimately judgment was entered for the subcontractor against the city. At that point, the court considered the general contractor's claims for the 14 percent and found that the markup could not be recovered. This court held that the markup for the general did not permit it to receive a portion of the subcontractor's judgment against the city based on the extra subcontract work. The pass-through agreement, however, permitted the general contractor to recover the attorneys' fees incurred in the suit against the city.

The amount recoverable as overhead under a change order, as provided for in the contract, was the subject of *Reliance Insurance Co. v. United States*, 20 Cl. Ct. 715 (1990). The contractor sought additional compensation in connection with a delay claim. The contractor was building an addition to a Veterans Administration hospital. During the course of the project, six separate claims arose. The contract provided:

> Allowances not to exceed 10 percent each for overhead and profit for the party performing the work will be based upon the value of labor, material and use of construction equipment required to accomplish the change. As the value of the change increases, a declining scale will be used in negotiating the percentage of overhead and profit. . . . Profit shall be computed by multiplying the profit percentage by the sum of the direct costs and computed overhead costs.

Overhead was defined as:

> Overhead and Contractor's fee percentages shall be considered to include insurance other than mentioned herein, field and office supervisors and assistants, watchman, use of small tools, incidental job burdens, and general home office expenses, and no separate allowances will be made therefore [sic]. Assistants to office supervisors includes all clerical, stenographic and general office help. Incidental job burdens include, but are not necessarily limited to, review and coordination, estimating and expediting relative to contract changes are associated with field and office supervision and are considered to be included in the contractor's overhead and/or fee percentage.

The court held that the contractual limitations on the recovery of overhead for change orders applied equally to claims for delay damages and for direct costs. Therefore, the court granted the government's motion for partial summary judgment.

§ 9.14 Types of Recoverable Costs

Page 268, add at end of section:

The plaintiff contractor sued the city for additional amounts arising out of the city's alleged change orders and delays allegedly caused by the city in *J.A. Jones Construction Co. v. City of New York*, 753 F. Supp. 497 (S.D.N.Y. 1990). The contract with the city incorporated federal regulations defining the calculation of payments for change orders and delays. The plaintiff claimed that this incorporation provided the federal courts with jurisdiction over such a dispute. The project was funded by a grant from the Environmental Protection Agency under the Clean Water Act and involved the construction of publicly owned water treatment plants. The court granted the city's motion to dismiss for lack of jurisdiction, holding that the incorporation of federal procurement regulations into the contract was

§ 9.15   LABOR COSTS

insufficient to support federal jurisdiction. The court found that the regulations did not result in a private right of action, since the regulations were only designed to protect the government's treasury.

§ 9.15 –Labor Costs

Page 268, add after first sentence of section:

In *Neal & Co. v. United States*, 19 Cl. Ct. 463 (1990), for example, the plaintiff government contractor brought suit to obtain an equitable adjustment in the contract price. The plaintiff had contracted to construct storage tanks at a naval base in Alaska, and it did complete the project. The concrete was to be precast with steel plates attached prior to shipment to the remote installation site. During the casting of the panels, it was discovered that the concrete bonded to the steel and made the panels bow excessively. The plans and specifications did not contemplate any bonding. The contractor notified the Navy and determined that the bowing would create additional stress, making installation difficult. The panels were numbered to permit panels with similar bowing to be erected next to each other. This required additional loading, off-loading, and installation work. This plan was adopted by the general contractor without protest from the Navy. The contractor submitted a bill for the services of a design consultant who would deal with any other problems caused by the bowing which might arise at the job site and thereby mitigate the need for extra work. The government disallowed this claim and the request for contract modification, on the basis that the contractor failed to establish that the bowing was due to a design deficiency. The court held that the contractor was entitled to an adjustment in the contract price. The court found that the bowing was due to defects in the government's design specifications. The court also found that the contractor's course of action with regard to the design and the additional costs was reasonable.

CHAPTER 10

TERMINATION CLAIMS FROM OWNER'S AND CONTRACTOR'S PERSPECTIVE

§ 10.2 Overview

Page 281, add to footnote 14:

The court was required to construe a termination-at-will clause in a subcontract in Desco Vitro Glaze v. Mechanical Constr. Corp., 159 A.D.2d 760, 552 N.Y.S.2d 185 (1990). The plaintiff was to furnish labor, materials, and equipment for the installation of seamless flooring called for in the general contract specifications. The plaintiff was told by the defendant that the specifications had been revised and the work called for in the subcontract eliminated. The subcontractor sued for expenses incurred and for lost profits. Under the general contract, the owner had the right to "abandon, postpone or terminate the work or any part thereof for any . . . reason." Upon such a termination, the general contractor was required to terminate any applicable subcontractors. The court affirmed the summary judgment dismissing the claim. The court held that the subcontract specifically stated that the termination-at-will clause in the general contract was applicable to bar the subcontractor's claim.

§ 10.17 –During the Project

Page 294, add to footnote 75:

A plaintiff subcontractor in L.K. Comstock & Co. v. United Eng'rs & Constructors, Inc., 880 F.2d 219 (9th Cir. 1989), was hired to install the electrical systems in the pollution control project under construction. The general contractor terminated the subcontract, claiming that the subcontractor failed to use its "best efforts" to accomplish the work by the completion date. The subcontractor sued the general contractor, claiming that the termination was wrongful. The general contractor counterclaimed for defective performance of the work under the subcontract. The Ninth Circuit, applying Arizona law, affirmed the lower court's judgment for the general contractor, finding that the general contractor had properly cancelled the contract and thus could recover any additional costs to reprocure the services.

§ 10.28 BASIC DAMAGE REMEDY

The subcontractor claimed that the contract terms were ambiguous with respect to formal charge orders. The court held that custom and usage between the parties served to waive the contractual provision, and that the subcontractor was estopped from relying on the formal procedures set forth in the contract. The fact that the general contractor failed to give the subcontractor 48 hours' notice prior to the termination was irrelevant, even though this was provided for in the contract, since the default could not be corrected within a 48-hour period. At the time of termination, the subcontractor was 12 weeks behind schedule and was not entitled to an extension of time.

Page 295, add to footnote 78:

In one case, a contractor sued an owner for breach of contract under a series of construction contracts. The contractor performed work on several McDonald's restaurants and claimed that the projects were wrongfully terminated by McDonald's. The contractor was obligated to advance payment to its subcontractors and then to submit sworn certifications of the payments to McDonald's for progress payments. The contracts provided that the contractor could be terminated for a breach of contract, in which case McDonald's would complete the work. The contractor was not able to pay its subcontractors without first being paid by McDonald's. It certified that the payments were made, while knowing that the certifications were false. The contractor contended that McDonald's knew of this practice and acquiesced. McDonald's denied such knowledge. The court, in Abrahamsen v. McDonald's Corp., 193 Ga. App. 868, 389 S.E.2d 386 (1989), affirmed the lower court summary judgment for McDonald's, thereby enforcing the terms of the contract even though the refusal to advance money to the contractor would make performance of the contract impossible. The contractor failed to present sufficient evidence of a parol agreement to waive the contract terms such as might, at a minimum, create a question of fact capable of defeating the summary judgment motion.

*

§ 10.28 Basic Damage Remedy

Page 308, add at end of section:

The adequacy of the claim submitted by a contractor was reviewed by the court in *Sun Cal, Inc. v. United States*, 21 Cl. Ct. 31 (1990). The government sought to dismiss the suit, claiming that the contractor failed to comply with the Contract Disputes Act. The contract involved the lease and construction of an office building. The government terminated the contractor for default, whereupon the plaintiff acquired the contractor's interest in the project and disputed the termination. The plaintiff filed a certified claim with the contracting officer, which was denied in its entirety. The court held that the vice president and chief financial officer of the contractor properly certified the claim under the statute, since that individual was the official responsible for completion of the contract. The claim consisted of the plaintiff's estimates. The use of such estimates was in good faith and was sufficient to satisfy the statutory requirements when certain cost components could not be established with absolute certainty at the time the claim was filed. The certification was made to the best of the officer's knowledge and belief and complied with the statute when that officer was in the best position to determine that the figures were accurate.

§ 10.29 – Before Work Begins

Page 308, add to footnote 147:

The plaintiff in Bildoc, Inc. v. Chicago Hous. Auth., 714 F. Supp. 317 (N.D. Ill. 1989), was the successful bidder on a public housing project, and had considerable experience in dealing with public housing authorities. The plaintiff claimed that the defendant housing authority breached the contract. The court held that the plaintiff had breached the contract by failing to procure a performance bond as required. This failure was a material breach of a condition contained in the contract. The bond was also required by state law. The defendant did not waive the failure to comply with the condition by not immediately declaring the contractor in default after the expiration of the 10-day period for obtaining the bond. The following provision was contained in the instructions to bidders:

> The failure of the successful bidder to execute such contract and to supply the required bonds within ten days after the prescribed forms [the Contract Documents] are presented for signature, or within such extended period as the [CHA] may grant based upon reasons determined adequate by the [CHA], shall constitute a default, and the [CHA] may either award the contract to the next responsible bidder or readvertise the contract. . . .

714 F. Supp. at 319.

In Ramirez Co. v. Housing Auth., 777 S.W.2d 167 (Tex. Ct. App. 1989), the developer sought to recover for work performed. The projects were to be developed on a turnkey basis. Under such an arrangement, the developer is only paid upon the completion of construction, under a contract of sale, when the completed project is sold to the authority. The developer is not entitled to recover preconstruction costs until there is an executed contract of sale. In such a situation, there may be a preliminary contract of sale wherein the authority agrees to purchase the drawings and specifications, but the developer may choose to eliminate this step under an accelerated turnkey basis.

The court in *Ramirez* held that there was a material question of fact precluding summary judgment for the authority, because there was evidence that the developer might have been entitled to the preliminary contract or to the execution of a full contract at the time the authority voted to terminate the developer. There was also a fact question as to whether quantum meruit was applicable. However, the

§ 10.30   DURING THE PROJECT

developer had no viable breach of contract claim with regard to a second project. The case clearly points out the risks of developing a project on a turnkey basis.

§ 10.30 –During the Project

Page 309, add to footnote 151:

A city terminated a general contractor's work after it was determined that the contractor had not obtained the necessary wetlands excavation permit. In Earthbank, Inc. v. City of N.Y., 145 Misc. 2d 937, 549 N.Y.S.2d 314 (1989), the contractor sued to recover amounts for additional work and for delay damages. The contract called for the contractor to be paid a unit price for excavation of the site, leveling, and refilling the site with fill that did not contain the weed *phragmites communis*. The contractor excavated thousands of cubic yards of soil in excess of the engineer's estimate before work was stopped due to the lack of a wetlands permit. The city ordered the contractor to restore the area. A change order was issued and the contractor was paid an additional sum for that work. The contractor sought to recover sums, in addition to the unit price, for other extra work, contending that the extra work was different in quality from that provided for in the contract. The court denied the city's motion to dismiss the contractor's claim for damages above the unit price. The court further held that the contract clause purporting to eliminate delay damages did not bar the contractor's delay damage claim, which was based on the city's alleged misrepresentation that it had obtained the necessary wetlands excavation permit.

In Martell Bros., Inc. v. Donbury, Inc., 577 A.2d 334 (Me. 1990), the painting subcontractor sued the general contractor for breach of contract when the general contractor told the subcontractor to leave the job. There were numerous delays arising from the failure of the buildings to be ready for the subcontractor's work. Structural problems also required that the subcontractor repaint certain areas. None of these delays and problems were due to any defective performance on the part of the plaintiff. The owner changed the type of paint, and the new paint permitted oil-based resin to bleed through. The type of paint and painting were again changed by the owner and general contractor to correct this problem. At a meeting, the general contractor sought to have the subcontractor do additional work without compensation. The contractor told the plaintiff that it "was losing money on the job and therefore [the Subcontractor] should accept the fact that [it] may lose money or break even on the job." *Id.* at 336. The relationship between the general contractor and the subcontractor further deteriorated as the entire project continued badly. There was a factual question as to whether the general contractor ordered the subcontractor off the job temporarily or permanently, and as to the reason for the order. The Maine Supreme Court affirmed the jury verdict for the subcontractor, holding that there was sufficient evidence to support a finding that the general contractor breached the contract. The court, however,

TERMINATION CLAIMS

held that the general contractor's statement about taking a loss was not an anticipatory breach of the contract.

§ 10.34 Other Damages Available

Page 315, add at end of section:

In *Nohcra Communications, Inc. v. AM Communications, Inc.*, 909 F.2d 1007 (7th Cir. 1990) (applying Pennsylvania law), a subcontractor sued the general contractor for breach of contract. The subcontractor was to dig ditches for the installation of an underground cable television system. The subcontract contained work schedules, specifications, and pricing details. Four months into the contract, the general contractor orally notified the subcontractor that it was terminating the agreement. The written notice for termination stated that the subcontractor failed to construct the required distance of ditch per day, failed to install cable in continuity, generated excessive homeowner complaints, and lacked experienced personnel. The general contractor failed to pay the majority of the subcontractor's invoices, whereupon suit was filed. The court held that the trial court's finding that the subcontractor performed in a commercially reasonable manner was supported by the evidence. There was also sufficient evidence to support a finding that the termination was not made in good faith. Clearly, the subcontractor could recover under its invoices for work performed. However, the court held that the subcontractor was not entitled to recover for lost profits, because the contract did not guarantee the subcontractor a certain level of work. Therefore, lost profits were highly speculative.

§ 10.35 Offsets Against the Contractor

Page 316, add to footnote 182:

In Thomas v. O'Brien, 791 S.W.2d 4 (Mo. Ct. App. 1990), the owner hired a contractor to build an addition containing a swimming pool and deck. The contractor had prepared architectural drawings and an estimate for materials and labor which he entitled "Bid Proposal." Subsequently, a document entitled "Contract Agreement" was signed by the parties. The contract called for payment in installments. The work was performed as agreed and the contractor performed the general construction work. When construction was 90% complete, the owner determined that the cost of labor and materials exceeded her expectations, and she terminated the contractor. She hired two carpenters who completed the work in three weeks for $1,605. The contractor sued for the balance of $10,000 remaining unpaid under the contract. The owner claimed that she had only agreed to pay the amount of the estimate. The trial court entered judgment for the contractor in the amount of the remaining balance, less the amount paid to the carpenters to complete the work. The appellate court affirmed, holding that the estimate was

§ 10.40 CONTRACTOR IN DEFAULT

only an estimate and not a contract. The bid proposal was not signed and did not evidence an intent to be bound. The document entitled "Contract Agreement" was binding and entitled the contractor to the four $20,000 installments, even though this was greater than the estimate amount of $70,870.

A plaintiff roofer sued a general contractor for breach of contract. The general contractor attempted to offset certain amounts claimed to be attributable to the failure of the plaintiff to fully perform its contract. The court, in Jerry B. Wilson Roofing & Painting, Inc. v. Jobco-E.R. Kelly Assocs., Inc., 151 A.D.2d 896, 542 N.Y.S.2d 867 (1989), held that no offset was available where the general contractor had not allowed the plaintiff to cure any claimed defects and the general contractor had excluded the plaintiff from the job site, thereby resulting in the incomplete performance.

§ 10.40 Contractor's Recovery When It Is in Default

Page 320, add to footnote 206:

In Timberland Paving & Constr. Co. v. United States, 18 Cl. Ct. 129 (1989), the plaintiff contractor sued to recover sums allegedly improperly withheld by the government under a road construction contract. The work included the excavation of unclassified material along a stretch of road and the use of the material as fill for several grade changes. The contractor was to change the road's vertical grade in several places and to replace the gravel surface with asphalt. The contract called for excavation by drilling and shooting presplit rock. The contractor investigated the site prior to bidding on the project.

The government officer authorized commencement of the work in October. Problems were encountered with early snow that caused delays in accomplishing the work. Work was then suspended for the winter. In the spring, work was started with only 170 days remaining on the contract. The contractor became involved in disputes with the subcontractor who was to haul and position the excavated material and was forced to replace that subcontractor.

The government issued a default termination to the contractor. The court held that the contractor was entitled to the remission of liquidated damages where the delay was due to unusual weather conditions; the government could not withhold certain compensation where the delay was due to such conditions. However, the court rejected the contractor's claim that several work stoppages or delays were due to safety problems and the actions of governmental agencies. Nor could the contractor recover additional costs incurred by having a replacement hauler, when the court rejected the contractor's claim that the replacement was for safety problems. However, the court found that certain amounts were improperly withheld by the government, and those amounts were subject to interest from the date the claim was submitted to the contracting officer.

CHAPTER 11

PAYMENT DELAY CLAIMS

§ 11.6 – AGC Subcontract Payments Clauses

Page 331, add to footnote 6:

In DEC Elec., Inc. v. Raphael Constr. Corp., 558 So. 2d 427 (Fla. 1990), the district court certified the following question to the Florida Supreme Court: "Must all payment provisions in contracts between contractors and subcontractors or suppliers that concern a condition or time of payment provision be construed as a matter of law?" *Id*. at 428. The subcontractor sued the general contractor for money due under the subcontract. The general contractor refused to pay, since it had not yet been paid by the owner, even though the subcontractor had satisfactorily performed its contractual obligations. The subcontract provided:

> No funds will be owed to the subcontractor unless the General Contractor is paid by the owner in accordance with the sworn statement. The subcontractor fully understands that in event of nonpayment by the owner to the General Contractor, the subcontractor has legal recourse against the owner through the Mechanics Lien Laws or other legal procedures for their [sic] correct monies due.

Id. The trial court upheld this provision in the contract as a matter of law, and was affirmed on appeal. The Florida Supreme Court reiterated the general precept that contract construction is to be made as a matter of law.

In OBS Co. v. Pace Constr. Corp., 558 So. 2d 404 (Fla. 1990), a subcontractor sought to recover for the labor and materials furnished in connection with the project. The general contractor did not pay the subcontractor because it had not been paid by the owner. The subcontract provided that payment to the subcontractor was not due until the subcontractor's work had been accepted and the general contractor received final payment from the owner for the subcontractor's work. The Florida Supreme Court held the general contractor liable to the subcontractor for payment, as the subcontract provisions did not shift the risk of nonpayment by the owner from the general contractor to the subcontractor. The general contractor's payment bond surety was also liable to the subcontractor for the amounts due.

In Bentley Constr. Dev. & Eng'g, Inc. v. All Phase Elec. & Maintenance, Inc., 562 So. 2d 800 (Fla. Dist. Ct. App. 1990), the subcontractor sued the general contractor and its surety for payment. The contractor claimed that the subcontractor delayed the completion of the project, and attempted to set off delay damages against the amounts owed to the subcontractor. The trial court entered judgment for the subcontractor. On appeal, the court held that the general contractor failed to prove the right to such a setoff. Therefore, the sums were

§ 11.6 AGC CLAUSES

due under the terms of the contract, which provided for progress payments within 10 working days after the general contractor received payment from the owner. The court found that this payment provision was ambiguous, and interpreted the clause to mean that payment was due within a reasonable time after the general contractor received the payment from the owner. The court recognized that such a provision shifts the risk of nonpayment between the general contractor and the subcontractor and noted that any ambiguities would be construed against the general contractor.

* In Crown Plastering Corp. v. Elite Associates, Inc., 166 A.D 2d 495, 560 N.Y.S.2d 694 (1990), the plaster subcontractor sued the general contractor and the surety on a courthouse construction project to recover for extra work performed and for the balance due under the subcontract. After the plaintiff substantially performed its contract, the county terminated the prime contract with the general contractor. The court affirmed the denial of the subcontractor's motion for summary judgment, because the subcontract provided that the receipt of payment by the general contractor from the owner was a condition precedent to final payment to the subcontractor. Specifically, the contract stated:

> Receipt of payment from the owner for the subcontractor's work is a condition precedent to payment by the Contractor to the Subcontractor. The Subcontractor hereby acknowledges that it relies on the credit of the Owner, not the Contractor, for payment of its work. . . . Final payment shall be payable to the Subcontractor thirty (30) days after the general contract work is completed and accepted upon the condition that final payment is received by the Contractor from the Owner and provided the Work described in this subcontract is fully completed and performed in accordance with the Contract Documents and is satisfactory to the Owner, Architect and Contractor.

* In another case, the plaintiff was the supplier of asphalt pursuant to a public contract. The asphalt was delivered according to the contract but the supplier was not fully paid. In Credit Gen. Ins. Co. v. Atlas Asphalt, Inc., 304 Ark. 522, 803 S.W.2d 903 (1991), the paving subcontractor ordered various paving materials from the plaintiff. On the day after the last delivery to the job site, the general contractor made a partial payment to the subcontractor. No portion of that payment was given to the supplier. The supplier subsequently sued the general contractor, the surety, and the subcontractor for the amounts owed. Under the surety bond, there was a limitations period which required that suit be filed within six months of the "final payment." The court held that the interim payment to the subcontractor was not "final payment" which started the running of the period. The fact that no subsequent payments were made did not convert that payment into final payment when other payments were still due. Therefore, the court affirmed the summary judgment for the supplier.

* In City of Corpus Christi v. Heldenfels Bros., Inc., 802 S.W.2d 35 (Tex. Ct. App. 1990), a subcontractor supplied beams for incorporation into a city recreation center. The subcontractor sued for payment. The city's contract with the general contractor specified a lump-sum payment and that the general would provide

standard performance and payment bonds pursuant to Tex. Rev. Civ. Stat. Ann. art. 5160. The general contractor provided documents which on their face appeared to be the bonds and were accepted by the city. However, after the general abandoned the project, the city determined that the bonding company did not exist and that the bonds were fraudulent. The plaintiff provided the beams and completed its duties under the subcontract. The city made monthly payments to the general contractor. The monthly statements of the general contractor listed $32,500 as the amount allocated to the beams. The city authorized payment to the general of $29,250 for the beams, which did not include the retainage. The city discovered cracks in the beams, and pursuant to the general contract, retained $20,000 to protect itself from the possibility of defects. Engineers found that the beams were not defective. In the intervening period, the general contractor abandoned the project and filed for bankruptcy. A new general contractor was hired to complete the work. The court reversed the judgment for the subcontractor, holding that a subcontractor must look to the general contractor with whom it is in privity for payment. The city was not on notice of any expectation on the part of the subcontractor that payment would be made directly to it, nor had the city undertaken to pay the subcontractor. The court further held that the subcontractor could not recover from the retainage.

§ 11.14 Calculating Interest for Commercial Contracts

Page 341, add at end of section:

In *State Department of Transportation v. American Insurance Co.*, 199 Ill. App. 3d 1068, 557 N.E.2d 932 (1990), a supplier sued the surety on its payment-performance bond. The plaintiff was only partially paid for rock and porous granular material supplied to a subcontractor on a public project. The plaintiff would bill the subcontractor on a monthly basis for materials delivered. Invoices were to be paid within 30 days of the billing, although there was no written agreement between the subcontractor and the plaintiff which gave the plaintiff purchase orders for the materials. At one point, the payments were delayed and past-due invoices accumulated. A payment was made to the subcontractor but failed to cover all the past-due invoices. The court held that an agreement by the plaintiff to accept only a portion of the amount due, after meeting with the subcontractor, did not prejudice the surety so as to absolve it of liability under its bond. Therefore, the plaintiff was entitled to recover under the bond and was also entitled to recover 5 percent prejudgment interest.

§ 11.16 Determining the Proper Interest Rate

Page 341, add to footnote 34:

In Unis. v. JTS Constructors/Managers, Inc., 541 So. 2d 278 (La. Ct. App. 1989), there were multiple contracts between the painting and insulation subcontractor

§ 11.17 FORESEEABILITY OF DAMAGES

and the general contractor for work on three apartment complexes. The subcontractor sought to recover sums allegedly due under the contracts, plus the penalty as provided for in La. Rev. Stat. Ann. § 9:2784(C), for the failure of the general contractor to pay after it received payment for the work from the owner. The trial court entered judgment for the subcontractor. The appellate court affirmed, holding that the general contractor was not warranted in withholding funds for the alleged failure of the subcontractor to perform within the time limitations. There was evidence that any such delay was due to the failure of other subcontractors to properly and promptly finish their work so that painting could be done. The general had ordered the subcontractor off one job, but was clearly still liable for the amounts due under the other contracts, and it could not offset or recover the sums it paid over the contract price to have another painting contractor finish the work on one project when that delay was not the terminated subcontractor's fault. The withholding of payment was without reasonable cause, and therefore it was proper to assess statutory penalties.

§ 11.17 Foreseeability of Other Payment Delay Damages

Page 343, add after first sentence of section:

In a Miller Act case, an excavation subcontractor sued a general contractor and its bonding company over the general contractor's alleged failure to pay the subcontractor for completed work through progress payments. The Miller Act was applicable, as the work was a government project involving the construction of two dam structures. The lower court judgment for the general contractor was affirmed on appeal in *Stewart v. C&C Excavating & Construction Co.*, 877 F.2d 711 (8th Cir. 1989). The court held that the refusal to make progress payments would justify the suspension of performance by the subcontractor. However, the amount in question was a payment for approximately $2,000. This was not a material breach of the contract, particularly where the general contractor was covering all equipment rental and payroll expenses. The subcontractor also sought additional sums for extra work. The subcontract specifically provided that any such additional sum would not be binding on the general contractor unless agreed to in writing by the general contractor or the government. The fact that the general contractor adopted the subcontractor's numbers, and passed them on to the government, was not the necessary agreement in writing. Therefore, the refusal to pay for additional work was not a breach of contract.

In *Sanchez Plumbing v. Aetna Casualty & Surety Co.*, 564 So. 2d 1302 (La. Ct. App. 1990), the plaintiff plumbing subcontractor sued the general contractor and its surety for nonpayment of sums due under the subcontract. The project was the construction of a cell block and other facilities at a state penitentiary. The project was accepted by the state and payment was made subject to a retained amount and with a deduction for delay damages. The plaintiff subcontractor filed a lien affidavit, but the amount secured by the lien was never paid. The trial court entered judgment for the plaintiff. The court held that the subcontractor could not be assessed any portion of the liquidated delay damages in the absence of evidence that the delay was due to the acts of the subcontractor. This issue was remanded. The court held that the general contractor was required to pay the subcontractor within a reasonable period after acceptance of the work by the state, without regard to the claims of the state against the contractor, and that more than a reasonable period had passed.

* A subcontractor sued the general contractor arising out of work to repair and replace certain roofs on state hospital buildings. In *Marshall Construction, Ltd. v. Coastal Sheet Metal & Roofing, Inc.*, 569 So. 2d 845 (Fla. Dist. Ct. App. 1990), various problems arose with respect to one roof which was defective and permitted water to penetrate the roof insulation. The subcontractor was not able to replace the roof and could not proceed unless it was paid by the general contractor for the other work which was completed. The subcontractor then stopped work on the project and the general contractor ordered it off the job and hired another contractor to complete the work. The plaintiff subcontractor sued the general contractor, claiming that it was wrongfully discharged and that the general breached an oral contract. The subcontractor further claimed that it was entitled to the benefit of the general contractor's performance bond. The general counterclaimed for breach of contract and failure to complete the work in a workmanlike manner. The court reversed the judgment entered for the subcontractor, holding that it failed to meet its burden of proving lost profits or costs incurred as part of its damages. The evidence supported a finding that the subcontractor breached the contract by failing to properly install the roof without further payment. The subcontractor failed to prove any breach of the oral contract.

CHAPTER 12

COST CLAIMS

§ 12.7 Was the Work Properly Billed as Cost-Plus Rather Than Under a Fixed Fee Term?

Page 363, add at end of page:

As part of repair work necessitated by a broken pipe, an insurer made arrangements with a contractor to perform the necessary work. The contractor brought suit, seeking a declaration of the amount of money to which it was entitled for the repair and restoration work. *Union Insurance Co. v. Bailey*, 234 Neb. 257, 450 N.W.2d 661 (1990). The question was presented as to whether the house had been totally destroyed, whether it had been completely repaired, and whether the insurer was liable up to the full policy limit. The agreement between the insurer and the contractor failed to state a contract price. This rendered the insurer liable for the reasonable value of the materials and labor. The court affirmed the lower court's judgment for less than the amount sought by the contractor, but in an amount determined to be the reasonable value of the materials and labor. The court held that the contractor was not entitled to recover costs, attorneys' fees, or prejudgment interest.

In *Vaccaro v. Smith*, 29 Ark. App. 175, 779 S.W.2d 193 (1989), a contract to construct a motorized trolley down a steep incline was allegedly breached by the owner, who failed to pay for the construction. The contract called for payment based on an hourly rate for the construction and installation work. There was an estimate of $8,000 which was clearly labeled as such. Upon completion of the work, the contractor submitted a bill for over $16,000. The owner refused to pay more than the $7,500 which he had paid during the course of construction. The contractor claimed that the nature of the terrain made an accurate estimate impossible. The owner claimed that the intent of the contract was that the work would cost $8,000. The court found that the use of the word *estimate* in the contract was not ambiguous and that the contractor was entitled to $8,000 plus the actual amount of expenses incurred above that amount, without any additional allowance for profit.

§ 12.14 Owner's Involvement in Cost-Plus Contracting

Page 369, add at end of page:

A contractor submitted a bid to do renovation work at the owner's home. After the bid in the amount of $55,300 was submitted, the contractor met with the owner to discuss proceeding on a cost-plus agreement. They reached an oral agreement to do the work, as set out in architectural plans provided by the owners, on a cost plus 10 percent overhead and 10 percent profit basis, subject to a cap in the amount of the original bid. On the first day of work, the owner told the contractor that he wished additional work and said he wanted to go cost-plus. The question was presented as to whether the job was then transformed into a straight cost-plus basis for all the work, or whether it was subject to the original cap and cost-plus for the extra work. No writing was ever made. The court, in *Ken's Construction Co. v. Liles*, 560 So. 2d 103 (La. Ct. App. 1990), held that the work was done on a straight cost-plus basis without the cap. The contractor was also entitled to interest on the amount due under the contract.

In *Gunter Hotel v. Buck*, 775 S.W.2d 689 (Tex. Ct. App. 1989), the plaintiff hotel sought a declaratory judgment to the effect that the architect had been fully paid under the contract. The plaintiff also alleged that the architect breached the contract and was negligent in the performance of his contractual duties. The architect was retained by the hotel in connection with a renovation and new construction project. The work actually included three separate projects. The architect was to receive compensation based on the construction costs in the amount of 7 percent for the renovation and 5.75 percent for the new construction. The amount paid on one portion of the new construction was lowered to 4.25 percent due to greatly increased construction costs. During the work, the hotel terminated the architect. The trial court entered judgment for the architect, finding that he was wrongfully terminated. The appellate court affirmed in part and reversed in part. The court held that the architect was entitled to the payment of his percentage on the basis of probable construction costs arising out of the design of the new construction. The amounts paid to a consulting structural engineer by the hotel were not to be deducted from the amounts due to the architect, since the architect had the right to challenge the amount of the engineer's bill where it was to be paid out of the architect's fee. The hotel paid a portion of the bill after the architect rejected it as unreasonable. The court held that the hotel had no right to terminate the architect where the architect had substantially performed according to the contract's terms and it was not his fault that the project had been permanently abandoned.

A plaintiff engineer was hired to perform professional services with regard to the conversion of a shopping center into condominiums. The court, in *Criswell v. European Crossroads Shopping Center, Ltd.*, 792 S.W.2d 945 (Tex. 1990), reversed the lower court judgment for the owners. The case centered around the terms of the compensation. The contract provided that the engineer would be paid his fee based on the sale price of the buildings "or as a whole project." The court held that this provision entitled the plaintiff engineer to recover his fee of 1 percent regardless of whether the buildings were actually sold. The right to payment was triggered by the transfer of the property by a contract for deed.

CHAPTER 13
CAUSES OF ACTION AND POSSIBLE RECOVERIES

§ 13.2 Breach of Warranties

Page 376, add at end of section:

In *Massachusetts Bay Transportation Authority v. United States*, 21 Cl. Ct. 252 (1990), the authority had an agreement with the Federal Railroad Administration (FRA) for a jointly funded renovation of an intercity rail passenger depot. The administration provided the design documents which were allegedly defective and resulted in delays and cost overruns. The authority sued the federal government for the damages attributable to the faulty designs. The authority had no power to make design decisions or to supervise the design engineers retained by the FRA, although it awarded construction contracts after receiving the designs. Upon discovering the design defects, the authority submitted to the FRA its change orders and the contractor's claims, but the FRA withheld consent. The authority incurred over $3 million for the change orders necessary to correct the defective designs' deficiencies. The court held that the FRA breached its warranty that the designs were accurate and proper. The court also held that the defective design issue was ripe for decision and denied the government's motion to dismiss.

The plaintiff brought suit against an architectural firm to recover for the services it performed, whereupon the architectural firm brought a third-party claim against the contractor, claiming that the services were provided to the contractor and not to it. In *Korstad-Tebben, Inc. v. Pope Architects, Inc.*, 459 N.W.2d 565 (S.D. 1990), the plaintiff moved for judgment on the pleadings after all parties had answered. The basis for the motion was that the defendant admitted that it owed the plaintiff the amount stated in the complaint. The trial court granted the motion for judgment, rejecting the affidavit of the defendant which clarified the factual situation involved in the case. The supreme court reversed, holding that there were factual questions precluding a grant of the motion. The court recognized that the defendant complied with the notice pleading requirements and should not be penalized therefor. The defendant specifically requested in the pleadings that the plaintiff's recovery be limited to the proceeds of a foreclosure on the plaintiff's mechanic's lien. The court held that it was necessary to consider the nature of the contract between the parties and who had the obligation to pay the plaintiff for its work. Similarly, it was necessary to determine if there was privity between the plaintiff and the third-party defendants.

CAUSES OF ACTION

§ 13.5 Negligence

Page 377, add to footnote 13:

A contractor was hired to repair faulty sewer lines. It defrauded the city, which then sought to recover payments made from the contractor and its surety. The work was to be supervised by engineers who were retained to provide all necessary professional engineering services. The contractor had been hired to inspect, test, and seal any and all leaking joints between sections of sewer pipe. The contractor was required to keep logs of its inspection, testing, and repair work. The logs were to be certified by the engineers who were to supervise the work. The contractor allegedly falsified work logs to indicate that repairs had been performed when they had not. The court, in City of Houma v. Municipal & Indus. Pipe Serv., 884 F.2d 886 (5th Cir. 1989), applying Louisiana law, found that the engineering firm breached its contractual duties through its certification of testing and repair work which had not been accomplished; the engineering firm could be held liable therefor. The surety was also held liable for the fraudulent acts of the contractor under the terms of the performance bond. The engineering firm could be held liable to the surety since it was reasonable for the surety to have relied on the engineering firm to monitor the work.

Watson, Watson, Rutland v. Board of Educ., 559 So. 2d 168 (Ala. 1990), arose from damage caused when a roof leaked. The school board contended that the leakage and consequent damage were due to the failure of the roofing contractor to follow specifications. Suit was filed against the general contractor, the roofing subcontractor, the manufacturer of the roofing membrane, and the architects for the project. The board settled with the general contractor and the roofing manufacturer for $100,000. The trial court entered judgment against the architect for breach of contract but in favor of the architect on the negligence claim. The appellate court reversed the judgment on the contract claim and affirmed the judgment for the architect on the negligence claim. The court found that there was no expert testimony regarding the architect's breach of the duty of care in connection with inspections of the project, and therefore the architect could not be held liable. The school board failed to prove the essential elements of its case. The court also found that the negligence action against the architect was time-barred because it was filed more than two years after the plaintiff determined that it had a claim against the architect.

A contractor sought to recover sums due under a street paving contract in Bechtold Paving, Inc. v. City of Kenmare, 446 N.W.2d 19 (N.D. 1989). The city had retained an engineer to draw the plans and specifications for the project. The engineer authorized the contractor's change of materials, and approved and accepted the contractor's work despite the fact that there were streaks and ridges in the pavement. The city was not pleased with the quality of the work. It alleged that the project was not done in a workmanlike manner, and sought to recover damages in order to have the streets repaved. The court found that the deficiencies in the paving were due to the change in the materials, which were improper for

§ 13.5 NEGLIGENCE

the purpose, and the improper application of those materials to the road surface. However, since the engineer had approved the work, the contractor was entitled to the balance due under the contract. The court recognized that liability could be imposed on the engineer for negligence in permitting the substitution and the subsequent acceptance of the project. Damages could also be awarded against the engineer for the cost of repaving the street.

* In Southeast Consultants, Inc. v. O'Pry, 199 Ga. App. 125, 404 S.E.2d 299 (1991), the plaintiff homeowner sued the engineering firm and land surveyors for failing to properly perform percolation tests. The alleged negligence of the defendants resulted in noxious odors emanating from the house's septic tank. The problem manifested itself three months after the plaintiff moved into the new home. The court held that the plaintiff could recover for the alleged negligence without establishing that he was in privity with the defendants. The court noted that privity concepts are irrelevant in negligence actions when it is foreseeable that the plaintiff would be injured by the negligence. The court held that the maximum damages available to the plaintiff would be the amount of the market value of the home at the time of purchase. There could be no recovery for additional amounts when the value of the house appreciated from the date of the sale to the date of trial.

* In Shook of W. Va. v. York City Sewer Auth., 756 F. Supp. 848 (M.D. Pa. 1991), a contractor sued the sewer authority for uncompensated costs incurred during reconstruction of a sewage treatment facility. The contract required the plaintiff to renovate two sewage treatment "trains" and perform piping and mechanical work in phases to permit continuous operation of the facility. The city hired a project engineer and construction manager. The engineer/manager was given a wide range of responsibilities, including processing applications, making recommendations for payment, reviewing and approving shop drawings, making certifications, approving change orders, and rejecting allegedly defective work. The engineer/manager was given the initial authority to resolve disputes between the parties. The court was required to construe various dispute resolution provisions contained in the contract. It denied the sewer authority's motion to dismiss the suit for the failure of the contractor to submit the claims to the engineer/manager for resolution, holding that the provisions did not turn submission into a condition precedent to suit against the authority.

* In Fairbanks N. Star Borough v. Kandik Constr. Inc., 795 P.2d 793 (Alaska 1990), the plaintiff had contracted to construct a road for the borough. The defendants were the design firm which prepared allegedly defective plans and specifications and the borough which allegedly warranted the adequacy of those specifications. The borough had contracted with the design firm to prepare an environmental assessment, a feasibility study, and the initial plans for a subdivision. In another contract, the firm was hired to survey and design the roads for the subdivision. Based on the designs, the borough solicited bids for construction of roads. The construction contract was awarded to the plaintiff, which experienced difficulty in performing pursuant to the specifications. The plaintiff claimed that the plans failed to state accurately the amount of material to be excavated and

CAUSES OF ACTION

did not provide enough space to dispose of cleared debris. The design firm also allegedly failed to implement necessary changes to provide compensation to the plaintiff for the defective plans. The claim against the firm sounded in malpractice and ordinary negligence. The court reversed the judgment on the jury verdict against the borough, holding that various erroneous instructions had been given to the jury. Specifically, it was error to permit the plaintiff to present evidence of damages using a total cost approach without instructing the jury that the only recoverable damages were those flowing from the borough's purported breach of contract. It was also error to instruct the jury that the plaintiff could recover under a quantum meruit theory with respect to work that was clearly under the terms of the original contract with the borough. Finally, with respect to the claims against the design firm, it was error to instruct the jury that the borough could recover from the firm upon a showing that it was not responsible for the contractor's damages due to its own negligence.

* In Pensacola Executive House Condominium Ass'n v. Baskerville-Donovan Eng'rs, Inc., 566 So. 2d 850 (Fla. Dist. Ct. App. 1990), the plaintiffs sued the engineers for professional malpractice arising out of the defendants' preparation of an engineering report. The report was to evaluate the general mechanical, structural, and electrical condition of a complex prior to its conversion to condominiums. Such a report is required by Fla. Stat. ch. 718. The report was completed and submitted in May 1981. In June 1983, the plaintiffs filed an administrative complaint with the Bureau of Condominiums alleging that the report had been improperly prepared and misrepresented the actual condition of the roof. Suit was not filed against the defendant engineers until April 1986. The court reversed the summary judgment for the plaintiff based on the statute of limitations. The court found that the engineering defendants were professionals who were subject to the professional malpractice statute of limitation and therefore the two-year limitations period had passed prior to the filing of the suit.

* In Newark Beth Israel Medical Center v. Gruzen & Partners, 124 N.J. 357, 590 A.2d 1171 (1991), a hospital contracted with the defendant architects for the design for a new building. The building was to be designed so that more floors could be added, pursuant to a master plan for expansion in phases. Twelve years after the initial design and construction of the original building, the owner prepared to construct the remaining floors. However, it discovered that the already constructed base was not capable of safely supporting the additional stories. The hospital spent an additional $1 million to assure that the additional stories could be safely constructed in a manner consistent with the original plan. The hospital then sued the architects to recover those costs. The court held that the architects were entitled to summary judgment on the basis of the 10-year statute of repose (N.J. Stat. Ann. § 2A:14-1.1) which applies to services provided by such professionals.

* In Floor Craft Floor Covering, Inc. v. Parma Community Gen. Hosp. Ass'n, 54 Ohio St. 3d 1, 560 N.E.2d 206 (1990), the plaintiff flooring contractor sued the owner and the architect for damages resulting from the defect in the flooring which the plaintiff installed. The contract was a standard AIA form. The architects prepared the plans and specifications under a separate AIA contract with the

§ 13.9 BREACH OF WARRANTIES

owner. The plaintiff installed the floor pursuant to the owner's installation instructions. After installation of the resilient vinyl flooring, the plaintiff detected bubbles beginning to appear. Thereupon, the plaintiff performed additional work to identify and correct the problem. The additional work cost $5,000. The plaintiff claimed that the concrete floor on which the flooring was installed had not cured properly and that excess water in the concrete had reacted with the flooring adhesive. The court held that the plaintiff had no standing to sue the architects with respect to its claim of defective plans and specifications resulting in economic injury, because the plaintiff had no privity with the architects nor any substitute for privity.

§ 13.9 Breach of Warranties

Page 383, add after carryover paragraph:

In *Richmond v. Grabowski*, 781 P.2d 192 (Colo. Ct. App. 1989), a contractor and subcontractor were sued for fire damage allegedly caused by their negligence. The owner and the contractor used AIA Document A101, which was designed to be used with A201, although A201 was not used in this case. The contract provided:

> 17.3 Unless otherwise provided, the owner shall purchase and maintain property insurance upon the entire work at the site to the full insurable value thereof. This insurance shall include the interests of the Owner, the Contractor, [and] Subcontractors . . . in the work and shall insure against the perils of fire. . . .
> 17.5 The Owner shall file a copy of all policies with the Contractor before an exposure to loss may occur.
> 17.6 The Owner and Contractor waive all rights against each other for damages caused by fire or other perils to the extent covered by insurance obtained pursuant to this Article or any other property insurance applicable to the work. . . .

Id. at 194. The owner did not purchase insurance. He argued that the contractors waived this provision by commencing work prior to receiving copies of the policies, as required by provision 17.5. The court rejected the waiver argument, finding that the contractors did not waive their reliance on the owner's contractual obligation to purchase the insurance. The failure to purchase the insurance placed the owner in the position of an insurer. The court affirmed the lower court's summary judgment for the contractor and subcontractor enforcing the unambiguous terms of the contract.

In *S&T Construction Co. v. Harris*, 789 P.2d 640 (Okla. Ct. App. 1989), the plaintiff contractor sued the owner for breach of contract. The contractor built the structure pursuant to the owner's plans and specifications on the owner's property. The structure, as built, encroached on the property line. The court held that the contractor could not be held liable for the encroachment, and was not required to indemnify the owner for any damages sustained. The contractor was entitled to recover the full amount due under the contract. Such a holding was

CAUSES OF ACTION

supported by the evidence that the owner had surveyed the property and directed the contractor as to where to construct the structure.

* In *Woodruff v. Johnson*, 560 So. 2d 1040 (Ala. 1990), the plaintiff homeowner sued the home builder after she moved into a house which was almost completed. She discovered various deficiencies in the construction, including chipped exterior bricks, a crawlspace of incorrect size, rainwater falling into the fireplace, windows that failed to properly close, a leak in a bathroom, and peeling wallpaper, among others. The builder failed to correct the problems despite numerous requests. The plaintiff brought suit on fraud and breach of contract claims. The court affirmed the judgment entered on the jury verdict for the plaintiff, holding that there was sufficient evidence that the defendant breached the contract.

* In *Employers Insurance of v. Mississippi State Highway Commission*, 575 So. 2d 999 (Miss. 1990), the contractor's liability insurers paid a claim brought against the contractor by a party injured in an auto accident. The insurers then sued the state highway commission for its breach of the implied warranty that the plans and specifications followed by the contractor for road resurfacing would result in a safe highway. The contractor applied a resurfacing mixture which was designed and manufactured in accordance with the plans and specifications provided to the contractor by the commission. The commission raised the following clause of the contract as a defense:

> Third Party Beneficiary Clause:
> It is not intended by any of the provisions of any part of the contract to create the public or any member thereof a third party beneficiary hereunder, or to authorize anyone not a party to this contract to maintain a suit for personal injuries or property damage pursuant to the terms or provisions of this contract. The duties, obligations and responsibilities of the parties to this contract with respect to third parties shall remain as imposed by law.

The Mississippi Supreme Court affirmed the trial court's grant of summary judgment for the commission. The trial court held that this provision removed the possibility of a suit for breach of warranty. The supreme court held that the commission could not be held liable since the failure to provide proper specifications would be a tort claim from which the commission was protected by sovereign immunity.

* ## § 13.12 Negligence

Page 384, add at end of section:

The city approved a subdivision project. The subdivider defaulted and the subcontractors who worked on the project sued the city in an attempt to recover payment for work performed. In *Charlie Brown Construction v. Boulder City*, 106 Nev. 497, 797 P.2d 946 (1990), the subcontractors proceeded against the city on unjust enrichment claims arising out of improvements made to city property that was off the subdivision project site. The subdivider filed for bankruptcy and the plaintiff's liens were frustrated by the construction lender's priority deed of

§ 13.16 BREACH OF WARRANTIES

trust. As part of the city's approval of the project, it had required that the subdivider post a cash deposit with respect to the required offsite improvements. No payment bond was required to assure that subcontractors would be paid. The city accepted title to the offsite improvements and released the cash deposit despite the fact that the subcontractors had not been paid. The court reversed the summary judgment for the city, holding that the city violated its own ordinance requiring payment bonds to protect subcontractors. This left the plaintiffs with a viable negligence claim against the city. The court noted that although cities cannot normally be liable for failure to enforce ordinances, there is an exception when there is a special relationship with a plaintiff. The court noted that the damages sustained by these plaintiffs were foreseeable.

In *Sarvis v. Maida*, __ A.D.2d __ , 569 N.Y.S.2d 997 (1991), the plaintiff was a drywall finishing subcontractor who was injured when he fell through a stairwell opening. He sued the owner. The defendant owner worked on the addition and performed framing work and the cutting of an opening in the subfloor. The plaintiff claimed that the owners breached the obligations imposed by New York labor law. The court held that, although the statute does create a nondelegable duty owed to a contractor, it does not apply to owners of one- and two-family dwellings if the owner does not control the work. The court held that homeowner did not control the work so as to support the imposition of the statutory duty. The open stairwell was an obvious risk with regard to which the defendant was not required to take protective measures.

In *Rapp Construction Co. v. Jay Realty Co.*, 809 S.W.2d 490 (Tenn. Ct. App. 1991), a fire occurred at the job site and the contractor sued the owner for materials that the contractor had purchased and delivered to the job site for ultimate incorporation into the project. At the time of the fire, the materials in question had not been installed or incorporated into the project. The contractor contended that its contract with the owner required that the owner maintain insurance to protect the contractor from such a loss. The owner denied that there was such an obligation and the court agreed. The contract provision required that the owner obtain such insurance to cover damage to the contractor's *work*, which was defined as:

> The entire completed construction or the various separately identifiable parts thereof required to be furnished under the Contract Documents. Work is the result of performing services, furnishing labor and furnishing and incorporating materials and equipment into the construction, all as required by the Contract Documents.

The court held that the owner was not required to provide coverage until the materials were incorporated into the project.

§ 13.16 Breach of Warranties

Page 387, add to footnote 56:

Suit was brought against the builder of townhouses for negligence in the construction of the townhouses purchased by the plaintiffs, and under claims of a breach of the implied warranties of workmanship and habitability. The Wyoming Supreme Court, in Deisch v. Jay, 790 P.2d 1273 (Wyo. 1990), focused on the nature of the implied warranties. The townhouses had problems with excessive humidity and dampness in the basements, which resulted in the development of mold and mildew. The plaintiffs sought to have the defects corrected or to recover money damages to compensate for the loss of value of the property, for the loss of use of the basements, or for the removal and replacement of the basement floors. The trial court entered judgment for the homeowners.

On appeal, the court affirmed, holding that the evidence supported a finding that the implied warranties were breached. The evidence also supported the award of $1,980 to each homeowner, which was the amount of the diminution in value of the property due to the defect. The small amount of the damages was the result of a finding that the dampness was largely due to the high water table in the area and that the problem was likely caused by capillary action. Evidence was presented regarding the cost of installing a capillary break that would solve or greatly reduce the problem. One plaintiff was also entitled to recover for damage to his personal property, which he placed in the basement after the builder told him there was no problem, because he demonstrated that he attempted to mitigate his damages by using fans and dehumidifiers to reduce the problem.

In Norman Properties v. Bozeman, 557 So. 2d 1265 (Ala. 1990), the plaintiffs contracted with the defendant for the purchase of a new home built by the defendant. Numerous problems centered around the drainage on the lot and the heating system were discovered shortly after the plaintiffs took possession of the house. Later, problems relating to the workmanship in the entire house were discovered. The plaintiffs hired an engineering firm to inspect the house. The inspection detailed numerous instances of faulty workmanship, some of which raised serious safety concerns. The plaintiffs determined that the house was no longer habitable, because of the defects and sewage backups caused by the installation of pipes of inadequate size. The plaintiffs stopped making mortgage payments and the mortgagee foreclosed. The lender purchased the property for $47,832 and at the time of trial there was a contract for the sale of the house for $48,000. However, the potential purchasers were unaware of any structural defects. The trial court entered judgment on the jury verdict for the plaintiffs. The Alabama Supreme Court affirmed, holding that the fraud suit was timely filed when it was filed within two years after the discovery of the major structural defects. The house was purchased for $46,000 plus closing costs and repairs, but was essentially worthless due to the defects. This supported a damage award for $50,000, as it was not economically feasible to correct the structural defects.

In Eastover Corp. v. Martin Builders, 543 So. 2d 1358 (La. Ct. App. 1989), a plaintiff motel owner sued the general contractor, the project architect, the plumbing subcontractor, the electrical subcontractor, and various consultants to recover its costs in repairing a plumbing system installed nine years before the failure was discovered. The evidence established that there were insufficient pipe hangers and that the existing hangers were spaced too far apart, in violation of applicable building codes. The failure of the system was caused by the collapse

§ 13.16 BREACH OF WARRANTIES

of certain sewer piping which was improperly supported. The plaintiff raised a variety of theories, including breach of warranty of workmanship and improper design and inspection, among others. The court reversed the lower court's judgment for the plaintiff. The court held that acceptance of the work by an owner without any complaint works an estoppel on claims for defective workmanship. This estoppel was effective against the plaintiff, a subsequent purchaser of the motel who was subrogated to the rights of his predecessor. Knowledge of the defective use of hangers, either actual or constructive, was imputed to the architect who was an agent of the owner, and this, being a patent defect, was also subject to estoppel. In this situation, the architect was also a general partner of the original owner.

In Cigal v. Leader Dev. Corp., 408 Mass. 212, 557 N.E.2d 1119 (1990), condominium purchasers sued the developer and subcontractors for breach of contract and for failing to construct the property in a workmanlike manner. The plaintiffs used an engineering report detailing construction defects and deviations from the plans. The plaintiffs' claimed damages were in the form of potential assessments which would be required to correct the defects. The court held that the plaintiffs did not have standing to assert the construction defect claims. Such claims could be brought only by the condominium association.

* The homeowners sued the contractor for a breach of the construction contract and for breach of warranty. The contractor sued to foreclose on the purchase money mortgage that secured its payment on the contract. In Argentinis v. Gould, 23 Conn. App. 9, 579 A.2d 1078 (1990), the contract incorporated house plans and specifications created by architects and provided for express warranties. There was a warranty against water leakage into the house for a period of one year after closing. Despite the closing and the grant of a certificate of occupancy, the homeowners were not initially able to occupy the house, because many items of construction had not been corrected. The parties modified the terms of the purchase money mortgage securing final payment of the total sales price. The homeowners moved into the house two months after the completion date set forth in the contract, although several items had still not been completed. There was also a problem with poor water pressure and contaminated well water, as the well had been placed, without a permit, too close to a septic system and a sewer drain. The homeowners had no drinking water for three months. Other defects were also found. The basement flooded during rains within the warranty period, resulting in substantial damage. The contractor refused to repair the defects unless it was given an unconditional release for all defects and was given the right to determine which defects were its responsibility. The contractor then sought to foreclose on its mortgage. The court affirmed the judgment for the homeowner, finding that there was sufficient evidence supporting the judgment for compensatory and punitive damages. The contractor failed to perform the construction contract in a proper manner.

* In Clements v. Barnes, 197 Ga. App. 120, 397 S.E.2d 560 (1990), the homeowners sued the contractor over defects in the construction of their home. During the first year after construction, there was a separation and cracking in the brickwork around a bay window, cracks in a ceiling, water leaks, and other defects. The complaints were communicated to the contractor who failed to respond prior to the suit. The contractor acknowledged that he warranted the home for one year consistent with industry practice. The court affirmed the judgment on the jury verdict for the homeowners, finding that the verdict was supported by the evidence. The measure of damages was the cost of restoration and repair of the defects.

* In Fetzer v. Vishneski, 399 Pa. Super. 218, 582 A.2d 23 (1990), the owner sued to recover the cost of replacing skylights which leaked. Six months after moving into the new house, the plaintiff found that one of the house's eight skylights was leaking. Shortly thereafter, all eight were found to leak. The plaintiff called and wrote to the defendant on numerous occasions seeking to have the defendant correct the problem. The defendant failed to even attempt to repair the skylights and the plaintiff hired another contractor to perform the work. The contractors correcting the work determined that the skylights had been improperly installed and their plastic was ripped. According to the second contractor, the only way to correct the problem was to replace the skylights. This was done, whereupon the plaintiff brought suit for breach of contract and breach of the warranty of habitability. The court affirmed the judgment for the plaintiff on the warranty claim, holding that there is an implied warranty of habitability attendant to such a construction contract and that there was sufficient evidence to support the conclusion that the warranty had been breached. In this case, the court found that the measure of damages was the cost of replacing the skylights rather than a lesser amount representing the diminution in the fair market value of the house due to the problem.

* In Wright Way Constr. Co. v. Harlingen Mall Co., 799 S.W.2d 415 (Tex. Ct. App. 1990), the owner sued the contractor and its surety for breach of contract and breach of warranty with respect to the paving of a parking lot for the owner's shopping mall. The parking lot started to "ravel," that is, rocks, sand, and other material in the asphalt broke away from the surface. The problem was due to the use of improper materials and/or air pockets in the asphalt. The contract required that the contractor use a "reputable commercial laboratory" to develop a batch design and for quality control of the materials. The court reversed the judgment against the contractor and the surety, finding reversible error in the failure of the trial court to instruct the jury on the issue of whether the owner was estopped from raising its claims because it failed to conduct a batch design monitored by the owner's independent geotechnical engineering firm.

* In Atheraton Condominium Apartment-Owners Ass'n v. Blume Dev. Co., 115 Wash. 2d 506, 799 P.2d 250 (1990), a condominium association sued the builder, architect, and others over defective construction of two buildings. The project architect submitted the architectural plans for the project to the city pursuant to Uniform Building Code (UBC) § 301. The plans were reviewed by a registered licensed architect who was head of the city building department. This review found some violations of the UBC which were corrected by the project architect, whereupon the plans were approved. The construction of the buildings was completed and certificates of occupancy were issued. Two years later, the exterior walls of

§ 13.17 CONTRACTOR'S DEFENSES

the buildings began to crack and materials believed to have been stucco started to fall off. The contractor completed repairs without charge. Two years later, the problem recurred, but the contractor refused to make repairs. The plaintiffs repaired the building, at which time they learned that the material was not actually stucco but rather a less expensive, inferior stucco substitute which would deteriorate further over time. The architectural plans did not authorize the use of the substitute, which did not satisfy the UBC fire resistivity standards. At that time, the owners also discovered other defects and UBC fire safety violations. The owners sued the contractor, the architect, and the city building department. The trial court granted the defendants a summary judgment which resulted in the dismissal of almost all claims. The appellate court held that there was a viable claim for breach of the implied warranty of habitability against the contractor which was not susceptible to summary judgment arising out of the violation of fire safety standards. There were also factual questions with regard to the fraudulent concealment claim against the contractor which precluded summary judgment. The court held that the claims against the city were properly dismissed, as was the claim for negligent construction of the building. There was no viable negligence claim where there were no allegations of personal injury or property damage and the complaint only sought to recover economic damages.

* In Deisch v. Jay, 790 P.2d 1273 (Wyo. 1990), the plaintiff homeowners sued a builder for negligence in construction of a townhouse, breach of the warranty of workmanship, and breach of the warranty of fitness for habitation. The court was required to determine the nature and extent of the warranties. The trial court determined that the warranty of habitability is breached if a reasonable person finding the defect would conclude that there was a major impediment to habitation. The defendant constructed townhomes, one of which was sold to the plaintiffs. The townhomes suffered from excessive humidity and dampness problems in the basements, which resulted in the development of mold, mildew, and odors. The homeowners sought relief to correct the defects or for the money representing the loss of value in the townhomes due to the defect or from the removal and replacement of the basement floors and loss of use of the basements. The plaintiffs did not seek rescission of the contract for the purchase of the townhouse. The court affirmed the judgment for the homeowners, holding that there was a breach of the warranty of habitability and of workmanship. The court held that the proper measure of damages was the diminution in the value of the townhouse, which was determined to be $1,980.

§ 13.17 –Contractor's Defenses

* *Page 388, add to footnote 60:*

In Winford v. Webster Gravel & Asphalt, Inc., 571 So. 2d 802 (La. Ct. App. 1990), the defendant contractor orally agreed to construct a parking lot. The defendant subcontracted the project to the plaintiff to perform the surfacing work.

CAUSES OF ACTION

The owner provided no written plans or specifications and only required that the soil preparation be accomplished with lime and that an asphalt penetration seal be applied to the surface. The contractor and subcontractors expressed their concerns that the chosen method of construction would be inadequate, but the owner rejected suggestions that soil cement be used as a ground stabilizer. When the subcontractor mentioned the instability of the base in certain areas, the owner did nothing. After the surfacing was completed, the parking lot rapidly deteriorated. The plaintiff subcontractor sued the contractor and the owner for the amount due and recognition of a lien previously filed on the property. The court held that the contractor and the plaintiff subcontractor complied with the plans and specifications provided by the owner. The defective condition of the parking lot was due to the owner's plans and not any deficient workmanship or defective materials. The owner had no defense to the claims for payment; thus, the subcontractor was entitled to be paid for the work performed on the project.

In Hieb v. Opp, 458 N.W.2d 797 (S.D. 1990), the plaintiff installed a sewage system but was not paid by the defendant homeowner. The defendant counterclaimed that the plaintiff's performance was defective. The homeowner had drawn the preliminary plans for the house, which described the electrical and plumbing systems, with the aid of the contractor. The plans were sent to an outside company for final drafting and preparation of materials. The homeowner then decided to act as his own general contractor and hired his own plumber and electrician. The plaintiff was hired to excavate the basement, install a sewage system, and bring a water line into the house. Due to the depth of the basement, the original system was no longer suitable, so the defendant homeowner made the decision as to which system would be substituted. About a week after the homeowner moved into the house, he experienced problems with the sewage system, which backed up into the basement. The plaintiff was unable to remedy the situation. A scientist from the state water department concluded that the system was improperly designed and constructed. The court affirmed the dismissal of both the suit and the counterclaim. The court held that the homeowner bore the risk of such a problem when he acted as his own designer and general contractor. As the system violated numerous regulations, the court held that the contract was void as a matter of law, because the agreement required that the plaintiff ignore these regulations.

Page 388, add to footnote 62:

In Village of Pawnee v. Azarelli Constr. Co., 183 Ill. App. 3d 998, 539 N.E.2d 895 (1989), the municipality sued the sewer contractor for allegedly defective construction of the village's sewer system. The basic problem related to an excessive flow of water through the system. The contractor brought the engineer and subcontractors into the suit in an attempt to allocate liability among the various parties. The court held that the portions of the work that were accepted by the city's engineer could not support a recovery against the defendant contractor. However, as to other portions that were accepted but contained latent defects that could not be detected upon an initial inspection, recovery could be had.

§ 13.17 CONTRACTOR'S DEFENSES

The contractor in Bechtold Paving, Inc. v. City of Kenmare, 446 N.W.2d 19 (N.D. 1989), sought to recover sums due under a street paving contract. The city had retained an engineer to draw the plans and specifications for the project. The engineer authorized the contractor's change of materials, and approved and accepted the contractor's work, despite the fact that there were streaks and ridges in the pavement. The city was not pleased with the quality of the work. It alleged that the project was not done in a workmanlike manner, and sought to recover damages in order to have the streets repaved. The court found that the deficiencies in the paving were due to the change in the materials, which were improper for the purpose, and the improper application of those materials to the road surface. However, since the engineer had approved the work, the contractor was entitled to the balance due under the contract. The court recognized that liability could be imposed on the engineer for negligence in permitting the substitution and the subsequent acceptance of the project. Damages could also be awarded against the engineer for the cost of repaving the street.

Page 388, add to footnote 64:

In City of Birmingham v. Cochrane Roofing & Metal Co., 547 So. 2d 1159 (Ala. 1989), the city sued various parties, claiming that the design and installation of a roof on the city airport terminal was defective. The roof was constructed in 1971 using what were then state-of-the-art materials and design. Due to the energy crisis at the time of the roof construction, the petroleum-based asphalt was of poor quality. The roof was completed in 1973 and was accepted by the architect after an inspection. After the inspection, the city accepted the building. Suit was filed in 1983, less than 50 days before the 10-year anniversary of the acceptance. The trial court granted summary judgment for the defendants based on the statute of limitations. The Alabama Supreme Court affirmed, holding that the cause of action accrues when performance is completed, because the claims essentially related to negligence or the failure to perform in a workmanlike manner. As respects the architect, the claim accrued upon the architect's acceptance of the project and was thus time barred, since it occurred more than six years from the date of the filing of suit. The six-year period also applied to the claims against the roofing contractor and barred the suit; even if the breach occurred upon the expiration of the two-year warranty, that breach was still outside the limitations period. The court rejected the city's argument that the limitations period was waived by the efforts of certain defendants to repair the roof, as at no time did the defendants conceal the defects, promise to repair in return for forebearance, or otherwise induce the city not to sue.

* *Page 389, add at end of section:*

A contractor installed a metal roof on the owner's warehouse. The roof was 250,000 square feet and was comprised of thousands of individual metal panels. The contractor experienced problems with misalignment of the panels during the roof installation. This was the cause of the dispute. The contractor filed a mechanic's lien after it was not paid, in response to which the owner counterclaimed that the roof was improperly constructed. The court in *Hermann v. Varco-Pruden Buildings*, 106 Nev. 569, 796 P.2d 590 (1990), set aside the finding of the trial court that the defects in the roof eaves were not functionally substantial. The court held that this finding was not supported by the evidence. The court, therefore, reduced the amount of the mechanic's lien by the amount necessary to correct the eave defects. The contractor also brought a claim for rain damage to insulation that it had brought to the job site but had not yet installed. The court held that the risk of such damage is on the contractor and not on the owner of the building.

§ 13.19 Negligence

Page 390, add to footnote 82:

A home purchaser sued the builder and the mortgagee which financed the work, in Hernandez v. Westoak Realty & Inv., Inc., 771 S.W.2d 876 (Mo. Ct. App. 1989). The plaintiffs claimed that the contractor failed to build the home shell in a workmanlike manner. The plaintiffs were acting as their own general contractor, and assumed the responsibility for the excavation, foundation, and finishing work. They contracted with the defendant contractor to build the shell for $12,498. The plaintiffs left the area and, when they returned, they found that excavation and foundation work had been done without their authorization. The defendant contractor proceeded to build the shell and then stopped after the bulk of the shell was finished. The plaintiffs claimed that there were substantial defects in the shell that rendered the house uninhabitable. While the complaint spoke in terms of negligence on the part of the contractor, it essentially was a breach of contract action, with the negligence claims merely stating claims for a breach of the duty to construct the shell in a workmanlike manner. The court found that there was sufficient evidence to support the breach of contract claim. The plaintiffs were entitled to recover the cost of substitute housing, demolition of the defective shell, and estimates and appraisals for repair costs. Since the contractor and the mortgagee were separate entities, the contractor's breach of contract had no effect on the validity of the mortgage and the plaintiffs' liability under the note.

In Mine Creek Contractors, Inc. v. Grandstaff, 300 Ark. 516, 780 S.W.2d 543 (1989), the plaintiff gas-station owner sued a highway repair contractor for damages arising out of the defendant's failure to maintain proper access to the station during the construction process. During the construction process, traffic was detoured around the plaintiff's premises. For part of this time, potential customers had access to the station, but later in the project there was no access. The claim sounded in interference with the business relationship and other tort theories. The court affirmed the verdict for the plaintiff, and held that the apportionment of damages between those attributable to the construction and those attributable to other causes was for the jury to determine. The contractor was held liable for negligence in failing to maintain the access.

§ 13.19 NEGLIGENCE

In Domingue v. H&S Constr. Co., 546 So. 2d 913 (La. Ct. App. 1989), the plaintiff, a jaywalking pedestrian, was injured when she tripped and fell while crossing the street. She sued the contractor who had recently resurfaced the street and the city. The plaintiff claimed that the contractor was negligent in failing to remove heavy string guidelines, placed adjacent to the edge of the pavement, following completion of the work. The lines were used to guide paving machinery. The trial court assessed 60% of the fault to the plaintiff, 40% to the contractor, and no fault to the city. The appellate court affirmed in part and reversed in part. The court held that the leaving of the string did not pose an unreasonable risk of harm and that the string was open and obvious. The court also upheld a provision in the resurfacing contract which required that the contractor provide liability insurance for claims arising out of its own negligence and negligence of the city in supervising the contractor's work. Such a provision did not violate the statutory prohibition in La. Rev. Stat. Ann. § 38:2216(E), which would not permit a requirement that the contractor indemnify the city for the city's own negligence.

Both the contractor and the owner sought to recover damages, claiming that the other party breached the contract, in Mancorp, Inc. v. Culpepper, 781 S.W.2d 618 (Tex. Ct. App. 1989). The owner also claimed a breach of warranty and charged that the contractor engaged in deceptive trade practices. The contractor claimed that the building was complete and sought a final payment, while the owner claimed that the work was defective. The court held that the owner was entitled to recover double the actual damages that did not exceed $1,000, under the Deceptive Trade Practices Act (Tex. Bus. & Com. Code Ann. § 17.41 *et seq.*), and the balance of the actual damages for the contractor's failure of performance. The damages were for the cost of repairing the building. The contractor was not entitled to the balance under the contract, because the jury found that the contract work was not completed and the architect had not signed a certificate of completion.

* In Uhley v. Tapio Constr. Co., 573 So. 2d 390 (Fla. Dist. Ct. App. 1991), the plaintiff contractor sought to enforce its mechanic's lien. The owners contracted with the plaintiff to build two homes on separate parcels that they owned. The contract between the parties was oral and there was a dispute as to whether the amount due was based on the bid price or whether the houses were built on a cost-plus basis. The owners gave the contractor surveys setting forth benchmarks from which elevations were to be measured. The surveys contained "assigned elevations" and it was stated that the "elevations shown are based on assumed datum." The survey further stated "bench mark assumed." The houses were built at incorrect elevations, which resulted in various defects including a propensity to flooding. The court reversed the judgment awarding partial damages to the owners and remanded the case. The court held, however, that the contractor was required to make an independent determination of the proper elevations and that failure to do so supported the imposition of liability on the contractor.

* The plaintiff homeowner sued a contractor that installed the tile flooring because of repeated cracking of the flooring. In Pollitte v. Sherman, 168 A.D.2d 761, 563 N.Y.S.2d 915 (1990), the tiling contractor also brought a third-party action against the general contractor. The homeowner's contract with the general provided allowances for certain items, including the flooring. The general claimed that it was unfamiliar with the tile flooring and then contracted with the subcontractor for the installation. The alleged cause of the repeated cracking was the failure of the general contractor to install appropriate subflooring. The court affirmed the judgment for the plaintiff against the subcontractor and the judgment for the subcontractor against the general contractor for contribution of one-half of the award to the plaintiff. The subcontractor could be held liable for a portion of the damages because it had represented that the subfloor was appropriate for the tile's installation.

* In May v. Ralph L. Dickerson Constr. Co., 560 So. 2d 729 (Miss. 1990), the original owners and subsequent owners of a building sued the general contractor for negligent construction of the building and for breach of implied warranties, including that of workmanship in the degree used by the construction industry generally. The subsequent owners brought the claim under the authority of Miss. Code Ann. § 11-7-20. The defendant countered with the argument that the claim was barred by the architect's signing of a certificate of substantial compliance, which effectively certified that the construction was consistent with the project specifications. The court held that the lack of privity of contract between the subsequent owners and the contractor was not a defense to the statutory claim. However, there were factual questions as to the effect of the architect's certificate which precluded summary judgment for the defendant. Therefore, the case was remanded.

* In Richardson v. Collier Bldg. Corp., 793 S.W.2d 366 (Mo. Ct. App. 1990), the general contractor and the subcontractors sued each other, with the subcontractors seeking payment for work performed and the general contractor claiming that it incurred additional expenses due to the failure of the subcontractors to properly perform pursuant to the contract. The general contractor was involved in other projects requiring the use of "engineered fill" and knew that the fill must be compacted and the compaction tested; if it were not properly compacted, problems could develop. The general hired a surveyor to prepare bid packages for soil contractors. The plaintiff subcontractor was ultimately given the contract. No detailed soil compaction specifications were prepared for the project, although the surveyor told the subcontractor that the fill was to be compacted to 95% optimum compaction. The other subcontractor was to perform soil compaction testing. The fill subcontractor had substantially completed its work and had not seen any engineers making soil compaction tests. The fill subcontractor had not been told that it was necessary to stop the work after each layer of fill was added and compacted so that testing could be performed. The subcontractor was allegedly only told by the general contractor's agent that things looked fine. The court held that the general contractor's onsite superintendent had the authority to correct problems and stop work and was the general contractor's agent. The agent's deposition testimony was admissible as a party admission. The fact that the general contractor continued construction operations, even after it learned that no soil compaction testing had been performed, barred the general contractor from recovering

§ 13.19 NEGLIGENCE

from the fill and testing subcontractors for the resulting damage to the building. The testing subcontractor did not breach its contract when the general contractor did not advise it when each layer of fill was added and failed to advise it of alternative testing methods after the general contractor discovered that all the fill had been added and compacted without testing. The fact that the fill was not fully compacted was excused by the fact that the general contractor failed to provide prompt testing. The court affirmed the judgment for the subcontractors.

* In City of Glens Falls v. Crandell Assoc. Architects, __ A.D.2d __ , 566 N.Y.S.2d 689 (1991), the plaintiff purchased a boiler from the defendant. The boiler was installed in the building and subsequently failed. The plaintiff alleged that the defendants negligently or in breach of contract specifications installed a boiler other than the model specified by the project architect. The court held that the claim against the plumbing contractor accrued, for statute of limitations purposes, on the date the installation was completed and the boiler was first put to its intended use. The court rejected the argument that the period started to run from the date when the plumbing contractor completed its work, which was after the boiler was in use.

* In Turner Constr., Inc. v. American States Ins. Co., 397 Pa. Super. 29, 579 A.2d 915 (1990), the general contractor sued the subcontractor's surety. The subcontractor was hired to perform excavating and backfilling work in connection with construction of a parking garage. The excavation work was properly performed; however, the backfilling work following completion of the foundation was allegedly improperly done. The subcontract contained various specifications for the backfilling work, including the compacting of materials. The general contractor raised questions as to the compacting on several occasions and the subcontractor was required to take corrective actions. The work was accepted by the general, but subsequently problems were detected, in the form of sidewalk subsidence due to improper compacting and backfilling. The general notified the subcontractor of the problem. After the subcontractor failed to respond, the general contractor brought a claim against the surety. The subcontract provided:

> The Subcontractor shall remove, replace and/or repair at its own expense and at the convenience of the Owner any faulty, defective or improper work, materials or equipment discovered within one (1) year from the date of the acceptance of the Project as a whole by the Architect and the Owner or for such longer period as may be provided in the Plans, Specifications, General Conditions. . . .

The court found that the project had been accepted by the owner on June 26, 1984 and that the subcontractor had notified the general contractor that it would not correct the problems on August 9, 1984. Suit was started June 21, 1985. The court affirmed the judgment for the general contractor, finding that the subcontractor failed to properly perform under the terms of the subcontract. The general's claim accrued on the date of the sub's refusal to correct the problem and not the earlier date when the problem was discovered.

CAUSES OF ACTION

§ 13.22 Breach of Warranties and UCC

Page 394, add at end of section:

In *Hillcrest Country Club v. N.D. Judds Co.*, 236 Neb. 233, 461 N.W.2d 55 (1990), the plaintiff filed two suits arising out of the failure of the roof of its clubhouse. In one suit, the plaintiff sued the installer of the roof, claiming that it failed to honor its warranties. The installer then brought a third-party action against the distributor that supplied the roof. There were claims against the manufacturer of the roof, the company which laminated the steel from which the roof was made, and the manufacturer of acrylic film which was used in laminating the steel. There were also breach of warranty claims against the distributor and manufacturer of the film. The parties treated the transactions as being for a sale of goods, which would fall within the UCC. There were issues as to the reliance of the plaintiff on various advertising materials provided by the defendants, which brochures also described the applicable warranties. The installer further warranted the work and agreed to correct any defective work. The court held that the judgment against the contractor on the express warranty claim was supported by the evidence. The statement by the contractor that the roof would last for 20 years was a "special warranty." However, the plaintiff failed to properly prove the amount of damages sustained due to the breach of warranty. The court rejected the stipulation between the parties to the effect that the plaintiff's damages were to be determined by the cost of repairing or replacing the roof. This stipulation dealt with a matter of law which could not bind the court.

In *American Aluminum Products Co. v. Binswanger Glass Co.*, 194 Ga. App. 703, 391 S.E.2d 688 (1990), the subcontractor sued the general contractor on an open account for the subcontractor's manufacture and installation of skylights for two construction projects. The subcontractor claimed that the amounts due were liquidated debts and that it was entitled to interest from the date the amounts were due. The general contractor counterclaimed, contending that the skylights were not leak-free and were not properly completed pursuant to the contract drawings and specifications. Further, the general contractor contended that it spent additional amounts to correct the defective performance by the subcontractor. The court held that the purchase orders from the general contractor constituted the contract between the parties, despite the fact that the order was different from the subcontractor's proposal. The purchase order was a counteroffer which was accepted by the subcontractor through its performance. The court found that the general contractor provided the subcontractor with substantial labor in connection with the projects, which labor was to be included under the purchase order. The court further found that the evidence supported a finding that the subcontractor did not correct the leaks in the skylights. Similarly, the subcontractor was liable for the cost of repairing the skylights. Although the subcontractor was entitled to recover the amount due on the contracts, that amount was more than offset by the general contractor's valid claims.

The owner sued a general contractor and a subcontractor, claiming that the contractor's defective work resulted in damage to the refrigeration system and

§ 13.23 SUPPLIER DEFENSES

concrete slab in an ice-skating rink. In *City of New York v. Kalisch-Jarcho, Inc.*, 161 A.D.2d 252, 554 N.Y.S.2d 900 (1990), the court affirmed the denial of the subcontractor's motion to dismiss, holding that the city could maintain a breach of contract action against the subcontractor when it demonstrated that it was an intended third-party beneficiary of the subcontract.

In *R.W. Kern, Inc. v. Circle Industries Corp.*, 158 A.D.2d 363, 551 N.Y.S.2d 218 (1990), the owner of a building sued a flooring contractor that provided defective tiles. The claim sounded in breach of warranty. The court was required to determine whether the four-year statute of limitations provided by UCC § 2-725 applied, in which case the action was time-barred, or whether the claim was actually one involving a service rather than a sale, in which case there could be no viable claim for breach of warranty. The plaintiff claimed that the contractor breached its contract by failing to deal fairly with the owner, when the contractor allegedly knew that the recommended tiles were so defective as to be inappropriate for the flooring job. The court reversed the dismissal of the complaint, finding that there were factual questions as to whether the contractor breached the contract.

In *Biomass One, Ltd. Partnership v. S-P Construction*, 103 Or. App. 521, 799 P.2d 152 (1990), the owner sued the general contractor for a breach of the construction contract and the general contractor then sued a subcontractor and its surety. There were various defects in the work manufactured and installed by the sub. The subcontract provided:

> Limitation of Time for Commencement of Claims—Sub-Contractor hereby agrees that any claim or question arising under the terms of this Contract, whether directly or indirectly, and as to which Sub-Contractor may wish to institute arbitration proceedings pursuant to the provisions of this contract shall be barred unless asserted by Sub-Contractor of [sic] the commencement of such arbitration proceedings pursuant to the terms of this contract within six months after any inaction the parties do further agree that any claim or cause of action of any kind arising out of or connected with this Contract, directly or indirectly, shall further be barred with respect to the institution of proceedings in a court of law unless asserted by the commencement of an action within one year after any inaction or omission or the occurrence of any matter to which such claim or cause of action relates.

The court affirmed the summary judgment granted to the subcontractor and its surety, holding that the one-year limitations period applied to claims whether brought by the general contractor or the subcontractor with the period commencing on the date the subcontractor discontinued work on the project. The general's complaint, filed more than a year after that date, was untimely.

§ 13.23 –Manufacturer/Supplier Defenses

Page 394, add to footnote 111:

In City of Birmingham v. Cochrane Roofing & Metal Co., 547 So. 2d 1159 (Ala. 1989), the city sued various parties, claiming that the design and installation of a roof on the city airport terminal was defective. The roof was constructed in 1971, using what were than state-of-the-art materials and design. Due to the energy crisis at the time of the roof construction, the petroleum-based asphalt was of poor quality. The roof was completed in 1973 and was accepted by the architect after an inspection. After the inspection, the city accepted the building. Suit was filed in 1983, less than 50 days before the 10-year anniversary of the acceptance. The trial court granted summary judgment for the defendants, based on the statute of limitations. The Alabama Supreme Court affirmed, holding that a cause of action accrues when performance is completed, as the claims essentially related to negligence or failure to perform in a workmanlike manner. As respects the architect, the claim accrued upon the architect's acceptance of the project, and was thus time barred because it was brought more than six years from the date of the filing of suit. The six-year period also applied to the claims against the roofing contractor and barred the suit, since, even if the breach occurred upon the expiration of the two-year warranty, that breach was still outside the limitations period. The court rejected the city's argument that the limitations period was waived by the efforts of certain defendants to repair the roof, as at no time did the defendants conceal the defects, promise to repair in return for forebearance, or otherwise induce the city not to sue.

§ 13.24 Negligence

Page 395, add after carryover paragraph:

The plaintiff homeowners sued a contractor and a concrete supplier to recover after the concrete slab upon which their house was built cracked and started to sink. The slab started to show signs of failure approximately six years after completion of the house. In *Troyer v. Webster Homes, Inc.*, 566 So. 2d 114 (La. Ct. App. 1990), the court held that the concrete supplier could be held 10 percent at fault for the failure of the slab, because it used excessive water in the concrete when the concrete was poured. This was supported by the evidence. Liability could be imposed without regard to whether the excess water was added to the concrete by the employees of the supplier on their own initiative or under the directions of others. Liability could be imposed because the employees knew or should have known that more water should not have been added and would result in weakened concrete. This was a breach of the supplier's duty of due care in the manufacture and supply of its product. The general contractor was also held liable for the damage sustained.

After plaintiff homeowner sued a contractor that installed tile flooring, due to repeated cracking of the flooring. In *Pollitte v. Sherman*, 168 A.D.2d 761, 563 N.Y.S.2d 915 (1990), the tiling contractor also brought a third-party action against the general contractor. The homeowner's contract with the general provided allowances for certain items, including the flooring. The general claimed that it was unfamiliar with the tile flooring and then contracted with the subcontractor for the installation. The alleged cause of the repeated cracking was the failure of

§ 13.27 TYPES OF BONDS

the general contractor to install appropriate subflooring. The court affirmed the judgment for the plaintiff against the subcontractor and the judgment for the subcontractor for contribution of one-half of the award to the plaintiff. The subcontractor could be held liable for a portion of the damages because it had represented that the subfloor was appropriate for the tile's installation.

§ 13.27 Types of Bonds

Page 396, add after first sentence of section:

A subcontractor declared bankruptcy. Thereafter, the general contractor sued the surety under the bankrupt's bid bond. The court granted the surety's motion to dismiss, holding that the surety had no duty to issue a payment or performance bond and that the general contractor generally acquiesced to the subcontractor's unsecured performance. This served to remove any liability of the surety, since the general contractor and the subcontractor had a general indemnity agreement. The surety had only assumed the obligation to pay any penalties, which would generally amount to 10 percent of the amount of the bid. Clearly, a different result would have ensued had there been a performance bond. The court rejected the contention that the general custom is for the bid bond surety to thereafter issue a payment and performance bond. Such a custom cannot impose a duty which is clearly contrary to the terms of the existing surety bond. Any attempt to impose a duty based on custom would also be void as contrary to the statute of frauds. *Chas. H. Tompkins Co. v. Lumbermens Mutual Casualty Co.*, 732 F. Supp. 1368 (E.D. Va. 1990).

Page 396, add after second sentence of section:

It should be noted that the characterization of a construction project as a public works project may have statutory implications and may affect the rights of subcontractors and materialmen. In one such case, the plaintiffs sought to recover amounts from the city by claiming that, as a public work, the city was required to provide a bond. The court, in *Judd Supply Co. v. Merchants & Manufacturers Insurance Co.*, 448 N.W.2d 895 (Minn. Ct. App. 1989), held that the project was privately owned and that the city only provided an economic development incentive involving tax increment financing and a land writedown conveyance. This municipal involvement did not transform the project into a public work, and therefore the statute mandating a contractor bond was inapplicable.

*   In another case in which the general contractor failed to pay the subcontractor the full amount owed on the subcontract, the subcontractor sued the District of Columbia and the general contractor. In *District of Columbia v. Campbell*, 560 A.2d 1295 (D.C. 1990), the subcontractor worked on a residential housing renovation. The subcontractor sought to hold the District liable for the general contractor's obligation when the District did not require that the general obtain a payment bond for the project. Such bonds are required under the D.C. Little

CAUSES OF ACTION

Miller Act, D.C. Code § 1-1104. The court held that the subcontractor failed to comply with the terms of the Little Miller Act in that it did not meet the statutory notice requirements which are conditions for a suit against the District. The court rejected the subcontractor's argument that the District could be held liable under a third-party beneficiary theory with respect to the District's contract with the general contractor and the District's failure to assure that the general contractor obtained the required bond. However, with respect to the subcontractor's claim against the general contractor, the court found that the jury was to decide whether there was an oral contract between the general contractor and the subcontractor for the work on the renovation project.

Page 397, add after carryover sentence:

A contractor was hired to repair faulty sewer lines. It defrauded the city, which sought to recover payments from the contractor and its surety. The work was to be supervised by engineers who were retained to provide all necessary professional engineering services. The contractor had been hired to inspect, test, and seal any and all leaking joints between sections of sewer pipe. The contractor was required to keep logs of its inspection, testing, and repair work. The logs were to be certified by the engineers who were to supervise the work. The contractor allegedly falsified work logs to indicate that repairs had been performed when in fact they had not. The court, in *City of Houma v. Municipal & Industrial Pipe Service*, 884 F.2d 886 (5th Cir. 1989) (applying Louisiana law), found that the engineering firm breached its contractual duties through its certification of testing and repair work which had not been accomplished. The engineering firm could be held liable therefor. The surety was also held liable for the fraudulent acts of the contractor under the terms of the performance bond. The engineering firm could also be held liable to the surety, since it was reasonable for the surety to rely on the engineering firm to monitor the work.

* *Page 397, add at end of section:*

Halterman v. United States Fidelity & Guaranty Co., 269 Cal. Rptr. 363 (Ct. App. 1990) (not officially published), was a suit brought by a subcontractor against the general contractor's surety, based on a theory of bad faith. The general contractor purchased a payment bond from the defendant as required by the applicable statutory public contract procedures. The plaintiff supplied all of the labor and materials called for in the contract, and also supplied additional labor and materials pursuant to change orders, but was not fully paid. As against the surety, the plaintiff claimed that the surety breached a duty of good faith and fair dealing which was allegedly inherent in the bond. The court held that there is no viable bad faith claim in the surety bond situation. Bad faith is only available in insurance situations, not in the surety context.

A county was the obligee under a performance bond. The county assigned its rights under the bond to another contractor which agreed to complete the project and repair the bond principal's deficient work. In *Transdulles Centre Ltd. Partnership*

§ 13.27 TYPES OF BONDS

v. USX Corp., 761 F. Supp. 430 (E.D. Va. 1991), the issue was whether the assignment of the county's interests under the subdivision agreement and bond were valid. Specifically, the original developer and bond principal allegedly failed to properly construct certain drainage improvements. The plans for storm drainage were approved by the county and, as part of the approval for construction, the principal entered into an agreement with the county:

> To construct all physical improvements in accordance with said plat, plans, and profiles, and applicable provisions of the Subdivision and Zoning Ordinances, including, but not limited to, adequate storm drainage system both on the subdivided property and on adjacent properties as needed, the construction of streets and roads in accordance with current standards of the Department of Highways and Transportation and the submission of as-built plans for all such public improvements.

There was a performance bond securing the completion of the work as required. After inspections of the work in progress, the county informed the developer that the storm drainage system was not in compliance with county ordinances. The county informed the new owner of the property that it would not permit full development of the property until changes were made in the drainage system. The county then assigned the rights to the plaintiff, who agreed to make the changes. The court denied the surety and principal's motion for summary judgment and upheld the ability of the county to assign its rights as consideration for proper completion of the work covered by the bond.

In *Western Insulation Services, Inc. v. Central National Insurance Co.*, 460 N.W.2d 355 (Minn. Ct. App. 1990), a subcontractor brought suit under a payment and performance bond issued in connection with a construction project. The subcontractor submitted a bid in connection with reinsulation work following asbestos removal, but was not fully paid for its work. The issue was whether the subcontractor was entitled to selectively apply a credit from the general contractor to unbonded projects and then recover under this bond. The court held that the subcontractor was not entitled to selectively apply the credits after the general contractor made a payment on the subcontractor's general account. If there are no instructions from the payor as to how a payment is to be applied, the court held that the payment is to be applied to the oldest outstanding account. In this case, there were also questions as to the exact nature of the contractual obligation between the general and the subcontractor.

In *Travelers Indemnity Co. v. Hennepin County*, 918 F.2d 66 (8th Cir. 1990) (applying Minnesota law), an insurer sought to be relieved of obligations under its performance bond issued with respect to a public works project. The county occupied the building but there were continuing problems with water leakage, movement of retaining walls, cracking glass, and other defects in performance which prevented the county's acceptance of the architect's certification of substantial completion. The general contractor would not perform the necessary remedial work without additional payment. The county sued the general contractor but did not notify the surety until more than three years had passed. The surety claimed

CAUSES OF ACTION

that the suit was barred by the two-year statutory limitations period. The bond provided:

> This bond shall cover any and all warranty, guaranty or corrective periods provided by the contract documents and shall be in effect throughout the duration of such periods.

The court held that this provision did not serve to extend the performance surety's liability under the bond to cover this claim.

In *Board of Supervisors v. Sentry Insurance*, 239 Va. 622, 391 S.E.2d 273 (1990), the county brought suit under a contractor's performance bond after the contractor failed to meet its performance obligations under the contract. The surety raised the defense that the county failed to file suit within six months of its giving notice of the contractor's default as required by the language of the bond. The court reversed the summary judgment granted to the surety, holding that the provisions of the bond did not unambiguously create a contractual limitations period.

§ 13.32 Damages Generally

Page 401, add to footnote 136:

In Knowles v. Westbrook Builders, Ltd., 188 Ill. App. 3d 343, 544 N.E.2d 121 (1989), the plaintiff owners sued the contractor for failing to complete construction on time, in compliance with building codes, and in a workmanlike manner. This case involved a contract for the construction of a home. The contract provided a starting date and a completion date 150 days thereafter. The construction started two weeks late. The owners repeatedly agreed to extensions of time of almost four months, at which time they asked the city building inspector to look at the house. The inspection revealed that the house did not conform to code in a variety of areas. The court found that the contractor clearly breached the contract and failed to cure the defects in its performance. The court further found that the owners properly terminated the contractor and stopped progress payments. The owners, who never moved into the house, recovered the diminution in the value of the house caused by the substandard performance.

In another case, home purchasers sued the builder and the mortgagee financing the work. The plaintiffs claimed that the contractor failed to build the home shell in a workmanlike manner. The plaintiffs were acting as their own general contractor and assumed the responsibility for the excavation, foundation, and finishing work. They contracted with the defendant contractor to build the shell for $12,498. The plaintiffs left the area and, when they returned, they found that excavation and foundation work had been done without their authorization. The defendant contractor proceeded to build the shell and then stopped after the bulk of the shell was finished. The plaintiffs claimed that there were substantial defects in the shell that rendered the house uninhabitable. Although the complaint spoke in terms of

§ 13.32 DAMAGES GENERALLY

negligence on the part of the contractor, it essentially was a breach of contract action, with the negligence claims merely stating claims for a breach of the duty to construct the shell in a workmanlike manner. The court found that there was sufficient evidence to support the breach of contract claim. The plaintiffs were entitled to recover the cost of substitute housing, demolition of the defective shell, and estimates and appraisals for repair costs. Since the contractor and the mortgagee were separate entities, the contractor's breach of contract had no effect on the validity of the mortgage or the plaintiffs' liability under the note. Hernandez v. Westoak Realty & Inv. Inc., 771 S.W.2d 876 (Mo. Ct. App. 1989).

Douglass v. Liccardi Constr. Co., 386 Pa. Super. 292, 562 A.2d 913 (1989), was a homeowners' suit against a contractor, seeking damages for defective construction. The issue presented was whether the measure of damages is the diminution in value caused by the defect, or the greater cost of repair. The court affirmed a lower court judgment that awarded the owners damages in the amount of repair costs. The court held that such an award was not excessive and did not give the owners a windfall.

* In Swain v. Harvest States Coops., 469 N.W.2d 571 (N.D. 1991), the homeowner sued the contractor of the modular home. Several years after the home was built, it was sold by the original purchaser to the plaintiff. Five years after construction was completed, the plaintiffs discovered that a basement wall was warping. Subsequently, it was discovered that cracks were developing in the basement floor. Tests performed by the defendant found that the foundation did not meet minimum guidelines for wood basement foundations. Shortly thereafter, the defendant disclaimed any liability with respect to the problem. The plaintiff filed suit and at trial established that the problem was due to hydrostatic pressure and the failure of the drainage system. At the time of construction, the defendants were aware of the high water table and therefore recommended a wood basement foundation. The court held that the plaintiff was entitled to recover the cost of repairing the damages to the home and was entitled to prejudgment interest from the date of the defendant's disclaimer. Additionally, the plaintiff could recover consequential damages sustained in the form of expenses of moving and storing household goods and additional living expenses incurred while the repairs were being performed.

* In Garcia v. Kastner Farms, Inc., 789 S.W.2d 656 (Tex. Ct. App. 1990), the owner contracted with the general contractor for construction of an agricultural irrigation reservoir. The contract called for excavation of a large volume of clay to seal the reservoir walls. This was subcontracted to the plaintiff. The contract set payment at $1.40 per cubic yard of naturally compacted material hauled to the job site. The plaintiff began to haul clay to the construction site. However, the local police ticketed the trucks for overloading. The truck drivers refused to continue hauling the clay after they received citations, since at the $1.40 price to carry smaller loads was unprofitable. The owner complained that more clay was required. The truckers were unreliable and various shortages occurred. The owner offered to pay $1.75 per cubic yard if more trucks were brought in. After this price change, the hauling proceeded without incident. The appellate court held that the owner had not ratified the increase in the rate per cubic yard, which was the product of duress. There was testimony that the owner would never have offered such a high rate except for the fact that no other truckers were available after the project started. The owner only acquiesced after the truckers decided that they would no longer perform at the agreed-upon rate provided for in the written contract. However, the contractor was entitled to be paid on the basis of quantum meruit.

* In Deisch v. Jay, 790 P.2d 1273 (Wyo. 1990), the plaintiff homeowners sued the builder for negligence in construction of a townhouse, breach of the warranty of workmanship, and breach of the warranty of fitness for habitation. The court was required to determine the nature and extent of the warranties. The trial court determined that the warranty of habitability is breached if a reasonable person finding the defect would conclude that there was a major impediment to habitation. The defendant constructed townhomes, one of which was sold to the plaintiffs. The townhomes suffered from excessive humidity and dampness problems in the basements, which resulted in the development of mold, mildew, and odors. The owners sought relief to correct the defects or for the money representing the loss of value of the townhomes due to the defect or from the removal and replacement of the basement floors and loss of use of the basements. The plaintiffs did not seek rescission of the contract for purchase of the townhouse. The court affirmed the judgment for the homeowners, holding that there was a breach of the warranty of habitability and of workmanship. The court held that the proper measure of damages was the diminution in the value of the townhouse, which was determined to be $1,980.

§ 13.33 Recovery of the Cost of Corrective Work

Page 402, add to footnote 141:

The plaintiff in Lochrane Eng'g, Inc. v. Willingham Realgrowth Inv. Fund, Ltd., 552 So. 2d 228 (Fla. Dist. Ct. App. 1989), purchased housing units which had insufficient sewage disposal systems. Suit was filed against the developer, the general contractor, the engineer, and the septic tank contractor. The claim sounded in a breach of implied warranty, and the central issue was the measure of damages under such claims and the measure of damages for the claim against the engineer who designed the system. Three experts claimed that the problems could be remedied by connecting the system to the city sewage system, by adding a secondary tank and increasing the drain fields, or by aerating the effluent. The cost of the alternatives was respectively $112,000, $8,000, and $25,000. The trial court awarded $45,000, which represented $3,000 for maintenance so as to allow the system to function temporarily pending a solution, $25,000 for the installation of the aerobic system, and $17,000 for an engineering study to determine the feasibility of hooking up the units to the city sewer system.

The appellate court affirmed in part and reversed in part. The court found that the developer had breached the warranty of fitness for habitation, and that the

§ 13.33 COST OF CORRECTION

measure of damages was to be determined under that theory. There was also a breach of the warranty that the sewage disposal system was adequate. The court held that the proper measure of damages was the amount required to repair the condition that would give the plaintiff what he was entitled to under the contract. Accordingly, the court reversed the award of the amounts for the engineering study, as the plans for the units did not call for city sewer connections. The other amounts were recoverable. As for the liability of the engineer, the court held that the engineer was liable for the increased costs of additional construction which could have been saved had the design been proper and the construction undertaken in connection with the entire project. The court also granted the developer indemnity for any consequential damages suffered due to the negligence of the engineer.

In Fleming v. Urdl's Waterfall Creations, Inc., 549 So. 2d 1057 (Fla. Dist. Ct. App. 1989), the contractor built a waterfall on the owner's property and was not paid. The contractor filed a mechanic's lien. The owner refused to pay the total contract price plus extras because the waterfall did not operate properly. The contractor made numerous attempts to correct the problem, but was unsuccessful. The owner then hired another contractor who corrected the problem. The owner filed a counterclaim against the plaintiff contractor for the amount paid to correct the problem. The court reversed the judgment for the contractor, holding that the plaintiff could only be entitled to the contract price and extras less the amounts paid to correct the defective work. If the cost of the correction exceeded the balance due to the plaintiff, the owner would be entitled to recover attorney's fees; otherwise, fees would be awarded to the plaintiff contractor.

A plaintiff homeowner sued a contractor for breach of contract arising out of numerous defects in the contractor's work on a house addition. Specifically, there were water leaks that resulted in rot to the addition, the roof was sagging and leaked, and there was a question as to whether the whole addition had shifted off its foundation. The plaintiff introduced expert testimony detailing the construction defects. In Kohn v. Johnson, 565 So. 2d 165 (Ala. 1990), the court affirmed the trial court judgment on the jury verdict for the plaintiff. The court held that the measure of damages was not the difference in the value of the property if properly constructed as opposed to the actual value of the property in its defective state. Instead, the proper measure of damages was the cost of repair or replacement so as to make the addition conform to what had been bargained for in the contract.

In a case arising out of home improvements, suit was brought to recover the cost of correcting the contractor's defective workmanship. The contract involved the removal of the rear of the house and the construction of an addition housing a hot tub. The contract did not provide the details of construction, but scheduled payments of $7,000 at the time the contract was signed, $3,500 when the room was "dried in," and $2,000 upon completion and inspection. The first two payments were made. The contractor allegedly completed the room, but the building inspector did not give final approval. There were a variety of defects, so the owner refused to pay the remaining $2,000. The owner testified as to the cost of correcting the defective construction. The court held that the owner was entitled to recover such amounts, plus loss of use damages and attorneys' fees. There was sufficient evidence to support the award of damages and to support the finding that the contract was breached. Doughty v. Simpson, 190 Ga. App. 718, 380 S.E.2d 57 (1989).

In Salard v. Jim Walter Homes, Inc., 563 So. 2d 1327 (La. Ct. App. 1990), the plaintiff home purchasers brought suit to rescind a construction contract and to recover damages. They had agreed with the builder on a $32,900 contract price. As payment, the purchasers executed a promissory note providing for 10% interest. The purchasers signed a completion slip, despite the fact that certain work was not completed, and they moved into the house. The crux of their claim was defective workmanship and the failure to correct the defects. The trial court rescinded the contract. On appeal, the court reversed, holding that the the contractor was entitled to the purchase price of the home at the time work was substantially completed. This was true despite the fact that the performance was faulty. However, the purchasers were entitled to offset against the purchase price any amounts spent to correct the defective work. The contract would not be rescinded where the court found that the house was not totally uninhabitable.

In Anuszewski v. Jurevic, 566 A.2d 742 (Me. 1989), the contractor sued the owners for breach of contract; the owners claimed that the work performed was defective. The contract was for the construction of a home, and provided for progress payments. At the scheduled completion date, the house was only half-completed. Six months later, the owners discharged the contractor, who then sued the owners. The owners claimed that the work performed was defective, and sought delay damages. The court held that the owners could recover the amount necessary to correct the defective work. This amount could include charges representing the repairer's overhead and profit, where such amounts were customary.

Quate v. Caudle, 95 N.C. App. 80, 381 S.E.2d 842 (1989), arose out of a contract to construct a home. The contract called for the contractor to build a log house for $66,300. After two-and-one-half months, the house still had no roof, rafters, basement, flooring, insulation, or mechanical system. As of the date the contractor stopped construction, the plaintiff owner had paid over $35,500 to the contractor and its suppliers. The plaintiff spent $15,700 over the contract price to complete the house. Suit was brought under the contract and for unfair and deceptive practices arising out of the misrepresentation of the cost of construction. It was found that the contractor intentionally underestimated the cost of construction in order to obtain the contract. The plaintiff was entitled to recover the additional construction costs and the cost of borrowing additional money to pay for the construction, and those amounts were trebled as a penalty.

A contractor sought to recover sums due under a street paving contract in Bechtold Paving, Inc. v. City of Kenmare, 446 N.W.2d 19 (N.D. 1989). The city had retained an engineer to draw the plans and specifications for the project. The engineer authorized the contractor's change of materials, and approved and accepted the contractor's work, despite the fact that there were streaks and ridges in the pavement. The city was not pleased with the quality of the work. It alleged that the project was not done in a workmanlike manner, and sought to recover

§ 13.33 COST OF CORRECTION

damages in order to have the streets repaved. The court found that the deficiencies in the paving were due to the change in the materials, which were improper for the purpose, and the improper application of those materials to the road surface. However, since the engineer had approved the work, the contractor was entitled to the balance due under the contract. The court recognized that liability could be imposed on the engineer for negligence in permitting the substitution and for the subsequent acceptance of the project. Damages could also be awarded against the engineer for the cost of repaving the street.

The plaintiffs in Freeman v. Maple Point, Inc., 393 Pa. Super. 427, 574 A.2d 684 (1990), sued the contractor-developer for damages arising out of claimed defects in the house they purchased. The plaintiffs claimed that the grading of the lot and the content of the soil resulted in surface water problems after heavy rains. The defendant attempted to correct the problem by regarding the lot on four separate occasions. The claim went to arbitration, and the plaintiffs were awarded nearly $6,000. They appealed to the trial court and received a jury verdict of $45,785. The appellate court reversed, holding that the plaintiffs failed to prove their damages and that the amount of damages awarded by the jury shocked the conscience. The court refused to award this amount, which allegedly represented the cost of repairs, where it was disproportionate to the amount of the loss in value. The house was purchased for $95,900, so the award of $45,785 was clearly not the diminution in the value of the house caused by the defect. There was testimony that it might cost between $10,000 to $12,000 to correct the problem.

In Mancorp, Inc. v. Culpepper, 781 S.W.2d 618 (Tex. Ct. App. 1989), both the contractor and the owner sought to recover damages, claiming that the other party breached the contract. The owner also claimed a breach of warranty and charged that the contractor engaged in deceptive trade practices. The contractor claimed that the building was complete, and sought a final payment, but the owner claimed that the work was defective. The court held that the owner was entitled to recover double the actual damages which did not exceed $1,000, under the Deceptive Trade Practices Act (Tex. Bus. & Com. Code Ann. § 17.41 *et seq.*), and the balance of the actual damages for the contractor's failure of performance. The damages were for the cost of repairing the building. The contractor was not entitled to the balance under the contract, as the jury found that the contract work had not been completed and the architect had not signed a certificate of completion.

* In Clements v. Barnes, 197 Ga. App. 120, 397 S.E.2d 560 (1990), the homeowners sued the contractor over defects in the construction of their home. During the first year after construction, there was a separation and cracking in the brickwork around a bay window, cracks in a ceiling, water leaks, and other defects. The complaints were communicated to the contractor who failed to respond prior to the suit. The contractor acknowledged that it warranted the home for one year, consistent with industry practice. The court affirmed the judgment on the jury verdict for the homeowners, finding that the verdict was supported by the evidence. The measure of damages was the cost of restoration and repair of the defects.

CAUSES OF ACTION

* In American Aluminum Prods. Co. v. Binswanger Glass Co., 194 Ga. App. 703, 391 S.E.2d 688 (1990), the subcontractor sued the general contractor on an open account for the subcontractor's manufacture and installation of skylights for two construction projects. The subcontractor claimed that the amounts due were liquidated debts and that it was entitled to interest from the date the amounts were due. The general contractor counterclaimed, contending that the skylights were not leak-free and were not properly completed pursuant to the contract drawings and specifications. Further, the general contractor contended that it spent additional amounts to correct the defective performance by the subcontractor. The court held that the purchase orders from the general contractor constituted the contract between the parties, despite the fact that the order was different from the subcontractor's proposal. The purchase order was a counteroffer which was accepted by the subcontractor through its performance. The court found that the general contractor provided the subcontractor with substantial labor in connection with the projects, which labor was to be included under the purchase order. The court further found that the evidence supported a finding that the subcontractor did not correct the leaks in the skylights. Similarly, the subcontractor was liable for the cost of repairing the skylights. Although the subcontractor was entitled to recover the amount due on the contracts, that amount was more than offset by the general contractor's valid claims.

* The owner sued the swimming pool contractor for negligent construction of a pool in Mason v. Yontz, 102 N.C. App. 817, 403 S.E.2d 536 (1991). The parties contracted for the construction of a pool on the plaintiff's property. The plaintiff paid the contract price when the pool was 80% completed. When the pool was subsequently filled, it was found to leak and was not level. The defendant allegedly failed to repair the problems. The trial court entered judgment for the owner on the jury verdict. The appellate court ordered a new trial, holding that it was error to instruct the jury on the negligence measure of damages in such a case. The measure of damages in this case was the cost of repairs necessary to render the pool in compliance with the contract and its warranty, or the reduction in the value of the pool in its defective condition from the value of the pool if it had been properly constructed.

* In Fetzer v. Vishneski, 399 Pa. Super. 218, 582 A.2d 23 (1990), the owner sued to recover the cost of replacing skylights which leaked. Six months after moving into the new house, the plaintiff found that one of the house's eight skylights was leaking. Shortly thereafter, all eight were found to leak. The plaintiff called and wrote to the defendant on numerous occasions seeking to have the defendant correct the problem. The defendant failed to even attempt to repair the skylights and the plaintiff hired another contractor to perform the work. The contractor correcting the work determined that the skylights had been improperly installed and that their plastic was ripped. According to the second contractor, the only way to correct the problem was to replace the skylights. This was done, whereupon the plaintiff brought suit for breach of contract and breach of the warranty of habitability. The court affirmed the judgment for the plaintiff on the

§ 13.33 COST OF CORRECTION

warranty claim, holding that there is an implied warranty of habitability attendant to such a construction contract and there was sufficient evidence to support the conclusion that the warranty was breached. In this case, the court found that the measure of damages was the cost of replacing the skylights rather than not a lesser amount representing the diminution in the fair market value of the house due to the problem.

* In Wilhite v. Brownsville Concrete Co., 798 S.W.2d 772 (Tenn. Ct. App. 1990), the owner sued the contractor who installed a swimming pool, claiming that the installation was defective. The plaintiff sued under claims of breach of contract and breaches of express and implied warranties. The contractor claimed that the damage was due to the intervening acts of third parties. The trial court found that the pool was not properly constructed and hydrostatic pressure caused the pool to float out of the ground. The court affirmed the judgment for the plaintiff, holding that there was sufficient evidence to support a finding that the pool was defectively constructed. The measure of damages included the cost of repairing the pool, even if that amount was in excess of the original contract price to construct the pool.

Page 402, add to footnote 142:

A contractor allegedly failed to properly compact the soil prior to building the plaintiffs' home. The cost of repair work exceeded the value of the house in a repaired condition. The defendants had stipulated that the soil was defectively compacted, but contested the measure of damages. They claimed that the proper measure of damages would be the diminution of the value of the home due to the defect. The court, in Orndorff v. Christiana Community Builders, 217 Cal. App. 3d 683, 266 Cal. Rptr. 193 (1990), affirmed the trial court judgment for the homeowners awarding them the cost of the repair work. The court properly considered the testimony of the owners that they wished to remain in the home and repair it. The court held that the personal reasons for staying in the home were sufficient to support the assessment of damages in excess of the diminution in value. In its damaged condition, the home was worth $67,500. The value of the home were it not defective would have been $238,500, and the cost of repair was $243,539.

In one case, a condominium association sued the designers and contractors of the complex for defective construction. The issue before the court in Council of Unit Owners v. Freeman Assocs., 564 A.2d 357 (Del. Super. Ct. 1989), was the appropriate measure of damages. The court held that regardless of whether a claim is one for negligence or breach of contract, if the repair cost is not excessive when viewed in relation to the diminution in value of the property due to the defect, the cost of repair is to be awarded. Clearly, this may result in an appreciation in the value of the property, but such an effect is irrelevant. It is necessary, however, for a plaintiff to establish that the type of repairs and their cost are reasonable.

CAUSES OF ACTION

Page 403, add to footnote 143:

Homeowner plaintiffs contracted with the defendant for the purchase of a new home built by the defendant. Numerous problems were discovered shortly after the plaintiffs took possession of the house. These problems centered around the drainage on the lot and the heating system, and other problems relating to the workmanship in the entire house were discovered later. The plaintiffs hired an engineering firm to inspect the house. The inspection detailed numerous instances of faulty workmanship, some of which raised serious safety concerns. The plaintiffs determined that the house was no longer habitable because of the defects, coupled with sewage backups arising from the installation of pipes which were too small. The plaintiffs stopped making mortgage payments and the mortgagee foreclosed. The lender purchased the property for $47,832 and at the time of trial there was a contract for the sale of the house for $48,000. However, the potential purchasers were unaware of any structural defects. The trial court entered judgment on the jury verdict for the plaintiffs. The Alabama Supreme Court, in Norman Properties v. Bozeman, 557 So. 2d 1265 (Ala. 1990), affirmed, holding that the fraud suit was timely filed when it was filed within two years after the discovery of the major structural defects. The house was purchased for $46,000 plus closing costs and repairs, but was essentially worthless due to the defects. This supported a damage award for $50,000, as it was not economically feasible to correct the structural defects.

Page 403, add to footnote 144:

Suit was brought against a builder of townhouses for negligence in the construction of the townhouses purchased by the plaintiffs and under claims of a breach of the implied warranties of workmanship and habitability. The Wyoming Supreme Court, in Deisch v. Jay, 790 P.2d 1273 (Wyo. 1990), focused on the nature of the implied warranties. The townhouses had problems with excessive humidity and dampness in the basements, which resulted in the development of mold and mildew. The plaintiffs sought to have the defects corrected or to recover money damages to compensate for the loss of value of the properties, for the loss of use of the basements, or for the removal and replacement of the basement floors. The trial court entered judgment for the homeowners. On appeal, the court affirmed, holding that the evidence supported a finding that the implied warranties were breached. The evidence also supported the award of $1,980 to each homeowner as the amount of the diminution of the property value due to the defect. The small amount of the damages reflected the finding that the dampness was largely due to the high water table in the area and that the problem was likely caused by capillary action. Evidence was presented regarding the cost of installing a capillary break that would solve or greatly reduce the problem. One plaintiff was also entitled to recover for damage to his personal property, which he placed in the basement after the builder told him that there was no problem, when it was demonstrated that he attempted to mitigate his damages by using fans and dehumidifiers to reduce the problem.

§ 13.47   NOTICE PROVISIONS

§ 13.35   Recovery of the Cost to Complete

Page 405, add at end of section:

In *Pools Markets South, Inc. v. Coggins*, 195 Ga. App. 50, 392 S.E.2d 552 (1990), the owners contracted with the defendant to build a swimming pool within 30 working days. The defendant stopped work after numerous delays and disagreements. At the time the defendant stopped work, the plaintiffs had paid most of the contract price and then hired another contractor to complete the work. The pool was finally completed 19 months late. The plaintiffs were granted judgment slightly in excess of the amounts paid to the completing contractor, without a deduction for the amount still due under the original contract. The appellate court affirmed the award, holding that it was not so excessive as to require a reduction, and found that the award was supported by the evidence.

§ 13.37   Recovery of Land Value

Page 406, add at end of section:

In *Southeast Consultants, Inc. v. O'Pry*, 199 Ga. App. 125, 404 S.E.2d 299 (1991), the plaintiff homeowner sued the engineering firm and land surveyors for failing to properly perform percolation tests. The alleged negligence of the defendants resulted in noxious odors emanating from the house's septic tank. The problem manifested itself three months after the plaintiff moved into the new home. The court held that the plaintiff could recover for the alleged negligence without establishing that he was in privity with the defendants. The court noted that privity concepts are irrelevant in negligence actions when it was foreseeable that the plaintiff would be injured by the negligence. The court held that the maximum damages available to the plaintiff would be the market value of the home at the time of purchase. There could be no recovery for additional amounts when the value of the house appreciated from the date of sale to the date of trial.

§ 13.47   Compliance with Contractual Notice Provisions

Page 413, add at end of section:

In *Krugman & Fox Construction Corp. v. Elite Associates, Inc.*, 167 A.D.2d 514, 562 N.Y.S.2d 188 (1990), the surety claimed that the suit brought by a subcontractor on the payment bond was untimely. Prior to completion of the construction project, the owner (board of education) terminated its contract with the general contractor. At that time, the plaintiff subcontractor was owed money for labor and materials. The bond provided:

> No . . . action shall be commenced hereunder by any claimant: . . . b) After the expiration of one (1) year following the date on which Principal; ceased work on said Contract.

CAUSES OF ACTION

The termination of the general contractor occurred on October 14, 1987. The plaintiff continued to attempt to obtain the money due it until it filed this suit on January 10, 1989. The court affirmed the surety's motion to dismiss, upholding the limitations period set forth in the bond.

§ 13.48   Determining Actual Damages

Page 413, add at end of section:

In *Howard v. Jay*, 203 Ill. App. 3d 539, 561 N.E.2d 274 (1990), a contractor sued the purchaser of the house foundation to recover the contract price. The trial court entered judgment for the defendant, finding that the contractor had not substantially complied with the terms of the contract and could not recover under quantum meruit. The appellate court reversed. During construction of the foundation, the defendant asked the plaintiff for additional materials, including steel to reinforce the foundation walls and windows. The plaintiff was also asked to perform some plumbing work. The plaintiff billed the defendant for the contract price plus the amount of the extras. The defendant refused to make any payment under the contract, claiming that the work was not performed in a workmanlike manner since the foundation floor was not level and the walls of the foundation were cracked. The appellate court held that there was evidence that any cracks did not affect the function of the foundation and that the plaintiff was entitled to recover the value of the work performed.

§ 13.51   Recovery Despite No Damages for Delay Clause

Page 417, add at end of section:

In *State Highway Administration v. Greiner Engineering Sciences, Inc.*, 83 Md. App. 621, 577 A.2d 363 (1990), the contractor sued to recover damages for delays that occurred during the plaintiff's preparation of contract documents for a highway project. The plaintiff entered into a contract to perform final design services for the project, including the final bridge and road design and preparation of the contract plans, specifications, and documents for bid advertisement. The plaintiff was also to review shop drawings and any construction redesigns. The payment was to be on a cost-plus-fixed-fee basis subject to an overall maximum amount. The defendant rewrote the contract to include a no damages for delay clause which provided:

> The Consultant agrees to prosecute the work continuously and diligently and no charges or claims for damages shall be made by him for any delays or hindrances, from any cause whatsoever during the progress of any portion of the services specified in this Agreement. Such delays or hindrances, if any, may be compensated for by an extension of time for such reasonable period as the Department may decide. Time extensions will be granted only for excusable delays such as delays beyond the control and without the fault or negligence of the consultant.

§ 13.51 NO DAMAGES FOR DELAY CLAUSE

The court held that the clause was enforceable even though the delay was caused by funding problems that were outside the contemplation of the parties. The court also held that the clause was not unconscionable, even though the problems extended the duration of the work from 15 months to more than 6 years.

In *White Oak Corp. v. Department of Transportation*, 217 Conn. 281, 585 A.2d 1199 (1991), the contractor sued the department to recover the balance due under a contract for highway construction. Specifically, the contractor sought to recover its delay damages. The court was required to interpret a no damages for delay clause. The contract required that the contractor complete the work within 1,650 days from the starting date. If the contractor failed to complete the project within that time, the contract permitted the department to deduct $1,000 per day for each day the completion was after the specified date. The contractor completed the work 392 days late. After completion, the contractor invoked a contract provision which permitted it to seek a retroactive extension of the completion date due to delays beyond the contractor's control. The department extended the date by 305 days for excusable delay and "winter shutdown." This resulted in a payment deduction by the department of $87,000. An arbitration decision held that all the delay was caused by factors outside the contractor's control, such as the failure of a gas company to remove its gas lines and the delay caused by the department in furnishing controls for the construction of temporary railroad lines. The arbitration found that the delays also resulted in other damages to the contractor for increased labor and overhead costs. The arbitration further found that the department failed to make timely periodic payments pursuant to the contract and that payments were required for extra work and departmental overcharges to the contractor for gravel the contractor removed during the project. The trial court entered judgment for the contractor on most of the arbitration award. The appellate court reversed in part, holding that the contractor could not recover delay damages attributable to the failure of the gas company to perform in a timely manner. The court held that the no damages for delay clause was not enforceable with respect to claims for bad faith, willful or grossly negligent conduct, uncontemplated or unreasonable delay, or from delays due to one party's breach of a fundamental contractual undertaking. The court also awarded the contractor interest on the late periodic payments.

A contractor agreed to perform construction work at six public schools. The plaintiff contractor sought to obtain additional amounts in excess of the contract price as delay damages after encountering delays at all construction sites. The board of education awarded an additional $145,000 in compensation on the contracts and an additional amount for delay damages on a seventh project which was subsequently started. The authorizations for the additional payments were never acted upon by all the various authorities required to approve the payment. The court affirmed the dismissal of the contractor's action in *Manshul Construction Corp. v. Board of Education*, 160 A.D.2d 643, 559 N.Y.S.2d 260 (1990). The court held that even though the construction contract contained a no damage for delay exculpatory clause, the plaintiff had the burden of proving that the delays experienced were wholly unanticipated. In this case, the mere evidence of additional costs due to the delay did not meet this burden.

CAUSES OF ACTION

§ 13.53 Recovery of Consequential Costs and Damages

Page 418, add after third paragraph:

In *Ambassador Development Corp. v. Valdez*, 791 S.W.2d 612 (Tex. Ct. App. 1990), the parties entered into two contracts whereby the subcontractor would perform work for the general contractor on a construction project. The first contract was for the excavation and construction of building pads in accordance with plans and specifications. The second contract was for the completion of all concrete foundations, concrete paving, and retaining walls, including forms and rebar. The general contractor also requested that the subcontractor perform extra work. This was undisputed. The subcontractor was not paid in full for this additional work and filed suit to recover that amount and to foreclose on its mechanic's and materialman's liens filed against the property. The general contractor counterclaimed that the work performed by the subcontractor was defective and incomplete. The court affirmed the judgment for the subcontractor. The jury had answered "None" with respect to questions on the cost of the completion and correction of the work. This did not conflict with the jury's finding of substantial performance so as to support an attack on the verdict. Therefore, the subcontractor was entitled to recover the full balance due on the contract. The court also found that there was sufficient evidence to support the jury's verdict. The court affirmed the statutory mechanic's and materialman's liens but held that such liens could not include prejudgment interest. The subcontractor was entitled to the lien under the required retainage statute in the maximum amount of the required 10 percent retainage plus funds, determined under the fund-trapping method, that were paid after statutory notice was given.

CHAPTER 14

CONSTRUCTION DELAY CLAIMS

§ 14.5 Identifying the Causes of Delay

Page 432, add after carryover sentence:

The case of *Bolton Corp. v. T.A. Loving Co.*, 94 N.C. App. 392, 380 S.E.2d 796 (1989), involved multiple prime contractors and the interpretation of N.C. Gen. Stat. § 75-1.1. One prime contractor responsible for heating and ventilation sued the general contractor/"project expeditor." The plaintiff brought a variety of claims sounding in breach of contract, negligence, and bad faith. The court held that the plaintiff prime contractor had no viable negligence claim against the general contractor. The statute precluded such a claim but provided the plaintiff with a statutory cause of action. This was true even though there was no direct contractual relationship between the prime and the general. The contractors agreed that the determination of the project architect as to the responsibility for the delay would resolve the dispute. In such a situation, the decision of the architect will be treated as being prima facie correct. Clearly, in any such delay damage claim, the plaintiff must establish the amount of the damages with reasonable certainty, as was done in this case through expert testimony. Such damages may include extended overhead and similar costs.

In *City of Elmira v. Larry Walter, Inc.*, 150 A.D.2d 129, 546 N.Y.S.2d 183 (1989), the contractor walked off the job and was sued by the city. The construction contract provided for liquidated damages in the amount of $1,000 per day for each day the project was delayed beyond the scheduled completion date. The contract also provided that for each day the project was completed ahead of schedule, the contractor would receive an additional $1,000. The contractor was to receive progress payments certified by the consulting engineer. The engineer authorized the first four such payments, approved the fifth payment in part, and rejected the entire claim for the sixth, seventh, and eighth payments because the city believed it was being overcharged. There were also other outstanding issues involving changes in the plans. The contractor stopped work. The court recognized the enforceability of liquidated damage clauses, but held that such provisions were inoperative where the contractor abandoned the work prior to the scheduled date for completion. The court held that the city was entitled to recover the cost of completing the parking garage, less any amounts not already paid under the original contract.

* In *Troise v. United States*, 21 Cl. Ct. 48 (1990), the plaintiff contractor on a renovation project for the Air Force contested various contract adjustments and the imposition of liquidated damages. The contract was for the renovation of 10 blocks of housing units. It was the third part of a four-part project on which the contractor had already performed the second part. The contract included the installation of kitchen cabinets, heating, ventilating, and air conditioning equipment, new bathrooms, and repairs to walls and ceilings. After the Air Force received the bids, it amended certain specifications. One of the amendments changed the liquidated damage clause for delay damages. Another modification, regarding the kitchen cabinets, required that they meet certain standards. The plaintiff assumed that it could use the same cabinets supplied during the second part of the renovation and did not question the change, but the Air Force rejected the contractor's cabinet samples as failing to meet the specifications. There was a dispute as to the interpretation of the specifications and whether the plaintiff's cabinets actually complied. The contractor was also directed by the Air Force to replace nonconforming heating, ventilating, and air conditioning equipment despite the fact that the deviation from the specifications was minor.

* The government had placed a disclaimer in the contract regarding the accuracy of estimates: the contractor was charged with verifying the estimates in making its bid. In this case, there was a discrepancy between the estimate and the actual amount of wall surface that required repairs.

* On the government's motion for summary judgment, the court held that there were factual issues concerning the failure of the government to approve the kitchen cabinets which precluded a grant of summary judgment. There were also factual questions as to whether the contractor was entitled to an extension of time for performance and an equitable adjustment due to the severity of the weather. There were yet other factual issues on the question of whether the government was aware that its wall surface estimates were incorrect and whether the disclaimer in the contract was sufficient to protect it from the contractor's claims. The court, however, upheld the liquidated damages provision of the contract where the daily damage figure for delays was a reasonable estimate of the government's anticipated loss. The court also held that the government could assess such damages even prior to the expiration of the period for performance when the contractor failed to comply with the contract's progress schedule.

Page 432, add to end of paragraph numbered 4:

In *Timberland Paving & Construction Co. v. United States*, 18 Cl. Ct. 129 (1989), the plaintiff contractor sued to recover sums allegedly improperly withheld by the government under a road construction contract. The work included the excavation of unclassified material along a stretch of road and the use of the material as fill for several grade changes. The contractor was to change the road's vertical grade in several places and to replace the gravel surface with asphalt. The contract called for excavation using the drilling and shooting of presplit rock. The contractor investigated the site prior to bidding on the project.

The government officer authorized commencement of the work in October. Problems were encountered with early snow that caused delays in accomplishing the work. Work was then suspended for the winter. In the spring, work was

§ 14.5 IDENTIFYING CAUSES

started with only 170 days remaining on the contract. The contractor became involved in disputes with a subcontractor who was to haul and position the excavated material, and was forced to replace that subcontractor. The government issued a default termination to the contractor.

The court held that the contractor was entitled to the remission of liquidated damages where the delay was due to unusual weather conditions, and noted that the government could not withhold certain compensation where the delay was due to such conditions. However, the court rejected the contractor's claim that several work stoppages or delays were due to safety problems and the actions of governmental agencies. The contractor could not recover additional costs incurred by retaining a replacement hauler where the court rejected the contractor's claim that the replacement was for safety problems. However, the court found that certain amounts were improperly withheld by the government, and that such amounts were subject to interest from the date the claim was submitted to the contracting officer.

In *Weaver-Bailey Contractors, Inc. v. United States*, 19 Cl. Ct. 474 (1990), the contractor was awarded a contract for the construction of beaches, breakwaters, boat ramps, parking areas, and other recreational improvements on Arcadia Lake in Edmond, Oklahoma. The work largely entailed grading, cutting slopes, finishing, and constructing a breakwater (using loose rock) on the slopes bordering the lake. The Army Corps of Engineers' estimate was that it would be necessary to excavate 132,000 cubic yards of materials of varying composition. During the course of the work, it was determined that the estimate was off by 41 percent, and that 186,695 cubic yards would have to be excavated. The original rate of pay was $3.42 per cubic yard and a contract modification was issued to pay the contractor $3.29 per cubic yard for the additional 54,695 cubic yards. The original work schedule was disrupted by the additional excavation work and not all of the work could be completed before winter. Winter weather caused damage by erosion to some of the work, which thus required repair in the spring. The court held that the delay in the completion of the work was excusable under the facts of the case, as it arose from unforeseeable causes beyond the control of the contractor. Because of the differing site conditions, the contractor was entitled to an equitable adjustment of the contract price based on a 10 percent profit rate, since it was established that without the additional excavation work the project would have been completed early.

A plaintiff contractor sued a city over work performed in connection with the construction of a new sanitary sewage facility in *Stone v. City of Arcola*, 181 Ill. App. 3d 513, 536 N.E.2d 1329 (1989). The contractor was to start work within 10 days of the notice to proceed, and completion was to be within one year unless otherwise extended. The notice to proceed was issued following the submission of payment and performance bonds. The notice to proceed contained a completion date 10 days after that originally stated. The contractor sought an extension due to abnormal amounts of rainfall. In order to accommodate a request of the city, the contractor performed some work out of sequence. There were also other disputes over the adequacy of the plans and specifications. These problems were corrected by engineers in the field, but resulted in additional delays. With the scheduled date nearing, the contractor sought a 90-day extension, but was only granted 60 days and was reminded that the city would enforce the liquidated damge provision of $200 per day for delays beyond the extension. Under these facts and conflicting expert testimony, the trial court entered judgment for the contractor. The appellate court affirmed the judgment, holding that the liquidated damage provisions were reasonable and generally enforceable. The running of the delay period under that provision terminated when the project was 95 percent completed, with only minor repairs and finishing work remaining to be done. The contractor was entitled to recover damages in the form of increased costs sustained due to the delay that was attributable to the city. The court, however, rejected the city's counterclaim for damages arising out of the delayed completion of the project, as the city's actions caused the delay.

Page 433, add to end of paragraph numbered 7:

In *McDevitt & Street Co. v. Marriott Corp.*, 713 F. Supp. 906 (E.D. Va. 1989), the contractor sued the owner to recover amounts due under the construction contract. In response to that claim, the owner brought a counterclaim for delay damages. The contract provided that the contractor would complete construction so that the owner could start business within 330 days of the specified commencement date. Due to a variety of factors, the hotel was completed 19 weeks late, thereby triggering a penalty clause. The case revolved around the causes of the delay and whether additional work was requested by the owner.

The court held that the contractor could not recover additional amounts for extra work related to soil conditions, because the contract specifically placed such a risk on the contractor. Any delay due to adverse weather conditions was also to be borne by the contractor. The owner, however, had constructively agreed to pay for additional soil work. Nevertheless, the owner's denial of time extensions was proper, and could not support a claim for other additional costs incurred by the contractor in attempting to meet the contractual time schedule. The court evaluated a variety of claims relating to the delay and additional costs, but the contractor was not excused from the delay and was not entitled to increased compensation. The court held that delay damages are to be determined with regard to the rental value of a completed building or a reasonable return on investment. Under this measure of damages, the owner was awarded over $310,000 for the 132-day delay.

§ 14.7 Compensable and Noncompensable Delays

Page 435, add at end of section:

In *Placeway Construction Corp. v. United States*, 920 F.2d 903 (Fed. Cir. 1990), a contractor brought suit to recover amounts due under its contract with the government. The plaintiff contracted with the Coast Guard to build residential

§ 14.10 LIQUIDATED DAMAGES PROVISIONS

housing on Governors Island. After construction was completed, the plaintiff submitted a voucher to the contracting officer requesting payment of the remaining balance due under the contract. The balance was not paid. Several months later, the plaintiff made a written demand on the contracting officer for the contract balance, 31 adjustments for additional work, and for delay damages caused by the government. None of these claims was certified as required by the Contract Disputes Act. The contracting officer denied the claims, finding that the contractor failed to complete the contract "in a timely manner" and delayed the government from starting the work of two other contractors. The court held that the plaintiff could not maintain the suit for declaratory judgment, because there was no jurisdiction in the Claims Court, due to the failure of the plaintiff to comply with the Contract Disputes Act's certification requirements.

Additionally, the Claims Court has no jurisdiction over claims exceeding $50,000, and this plaintiff raised all its objections in a single claim. The appellate court held that the Claims Court erred in determining that there was a single claim that exceeded the jurisdictional amount. There was more than one set of operative facts and therefore more than one claim, for jurisdictional purposes.

After a contractor failed to meet the contractual schedule for completion of a rental unit complex, it abandoned the job. In *Arbor Club v. Omega Construction Co.*, 565 So. 2d 357 (Fla. Dist. Ct. App. 1990), the owner sued the contractor and the surety under its performance bond. The project was to be completed in phases; the initial units were to be rented while construction on subsequent units continued. The owner notified the surety after the contractor left the job. The surety exercised its option and found another contractor to complete the project. However, the surety refused to compensate the owner for damages caused by the delay in completion of the project. The court held that the bond provided coverage for delay damages and reversed the judgment on the jury verdict for the surety. The court found that the trial court erroneously instructed the jury that the surety was not liable for delay damages.

§ 14.10 Contracts with Liquidated Damages Provisions

Page 437, add to footnote 17:

The contract for construction of a shopping center provided for liquidated damages in the amount of $1,000 per day if the project were not substantially completed within 300 days of commencement. There would be a $1,000-per-day bonus for each day it was completed early. Within six months of the commencement of construction, the two anchor tenants were in their spaces. The remaining spaces were given occupancy certificates within the 300-day period. One anchor tenant in an outparcel was not given an occupancy permit until more than one year from the commencement of construction. At that point, the contractor walked off the job for nonpayment and sued the owner. The owner counterclaimed for liquidated damages. The trial court awarded judgment for the owner. The appellate court, in J.M. Beeson Co. v. Sartori, 553 So. 2d 180 (Fla. Dist. Ct. App. 1989), reversed and remanded, holding that the contract was substantially completed when the owner could place tenants in possession and collect rents. The court further held for the contractor by finding that it had just cause for stopping work when the owner failed to make payments pursuant to the contract. The case was remanded for a determination of the date when substantial completion occurred and the computation of damages.

Page 437, add to footnote 18:

Fred A. Arnold, Inc. v. United States, 18 Cl. Ct. 1 (1989), was an appeal by both parties from a decision of the Armed Services Board of Contract Appeals, arising out of the construction of military housing units. The contractor was seeking damages for government-caused delays and disruptions and for extra work. The government sought liquidated damages for a 13-month delay in project completion. The initial construction was completed in a timely manner, but the balconies required structural modifications that delayed the use of the housing units for 13 months. The board of appeals rejected almost all of the contractor's claims and also denied the government's claim for liquidated damages. There were substantial negotiations and a settlement was arranged. The board of appeals held that the settlement was an accord and satisfaction, but the contractor claimed that the consideration was inadequate. The court held that there was a valid accord and satisfaction which resolved most of the claims. The court found that the evidence supported the finding that the Navy did not cause unnecessary delays, and awarded liquidated damages pursuant to the contract provision, which was consistent with government regulations. Such liquidated damages were recoverable without proof that actual damages were suffered.

* *Page 437, add to footnote 21:*

In Southeast Alaska Constr. Co. v. State Dep't of Transp., 791 P.2d 339 (Alaska 1990), the plaintiff contractor was to perform airport runway improvements. The work was never completed and the contractor filed a claim against the state for additional compensation, whereupon the state brought a claim for liquidated damages. The design of the project contained grading errors and the stockpiles and embankment materials were either inadequate or unavailable. The project was substantially redesigned and its scope was expanded. After the contractor started work on the expansion project, the state issued a change order to set forth the new agreement, which the contractor refused to sign. The contractor submitted a claim for delay damages and extra work and sought an extension of the completion deadline. The deadline was extended and the contractor was granted additional amounts. Meanwhile, the surety became insolvent. The contractor filed a bankruptcy petition and the state moved to lift the automatic stay to permit termination of the contract. The state wanted to withhold payment on the contract until the plaintiff provided replacement bonds. The court affirmed the summary judgment for the state, holding that the plaintiff was only entitled to be compensated for

§ 14.13 CONSEQUENTIAL DAMAGES

the extra equipment at the rate used by the plaintiff in the bid for the original contract. The state was awarded liquidated damages for the failure to complete the project by the deadline under the terms of the contract. The court upheld the liquidated damages provision, as the damages were reasonable.

A debtor contractor sought to compel the city to release the retainage it held under two construction contracts and various change orders. In *In re* Florida Precast Concrete, Inc., 112 B.R. 451 (Bankr. M.D. Fla. 1990), the question was whether the retainage was the property of the debtor's estate. The court found that the debtor's work on the project was substantially completed on the date of the opening of the building. The court then considered the provisions in the contract for liquidated damages due to delays in completion, and was required to determine whether the liquidated damages provision was a penalty. The court upheld the $1,000-per-day provision, finding that it was not a penalty and was not unreasonable in view of the potential damages the city could sustain due to the delay. The court found that the contractor, with two other contractors, was responsible for the delay and therefore could only be held liable for one-third of the daily penalty up to the date when the city signed off on the punch list.

§ 14.13 –Consequential Damages

Page 440, add to footnote 31:

A contractor built two houses for the plaintiffs in Russo v. Heil Constr., Inc., 549 So. 2d 676 (Fla. Dist. Ct. App. 1989). The homeowners refused to pay the contractor. They claimed that the contractor abandoned work prior to completion, and sought delay damages. They attempted to establish the measure of damages through rental values. The proper measure of damages for the delay was found to be the fair rental value of the property being improved, and not the rental value of the property leased by the homeowners for substitute housing. The court further held that the homeowners could properly offset their delay damages from the amount due under the contract.

CHAPTER 15

CLAIMS FOR LOST PROFIT

§ 15.5 Calculation of Profits

Page 459, add to footnote 35:

In another case arising out of allegedly defective construction, the plaintiff sued the general contractor, which third-partied various subcontractors, some of their sureties, and the architect. The court, in Guy Williams Realty, Inc. v. Shamrock Constr. Co., 564 So. 2d 689 (La. Ct. App. 1990), held that the owner proved the defective condition of certain parts of the building. The proper measure of damages in such a situation is the cost of repairing or replacing the defective building. Additionally, the owner was entitled to recover lost profits where there was sufficient proof to establish the amounts with reasonable certainty. In this case, the evidence was introduced by means of a witness expert in the area of commercial property management. The lost profits were caused by the inability to lease portions of the building, due to persistent leaking in those areas.

TABLE OF CASES

Case	Book §
Abrahamsen v. McDonald's Corp., 193 Ga. App. 868, 389 S.E.2d 386 (1989)	§ 10.17
Acquisition Corp. v. American Cast Iron Co., 543 So. 2d 878 (Fla. Dist. Ct. App. 1989)	§ 9.9
Aetna Casualty & Sur. Co. v. Canam Steel Corp., 794 P.2d 1077 (Colo. Ct. App. 1990)	§ 1.5
A.G. Lichtenstein, P.E. v. Goldin, 166 A.D.2d 328, 560 N.Y.S.2d 780 (1990)	§ 9.5
Al Johnson Constr. Co. v. United States, 20 Cl. Ct. 184 (1990)	§ 8.22
Al Johnson Constr. Co. v. United States, 10 Cl. Ct. 732 (1990)	§ 9.5
Allen & O'Hara, Inc. v. Barrett Wrecking, Inc., 898 F.2d 512 (7th Cir. 1990)	§§ 5.6, 9.3
Ambassador Dev. Corp. v. Valdez, 791 S.W.2d 612 (Tex. Ct. App. 1990)	§ 13.53
American Aluminum Prods. Co. v. Binswanger Glass Co., 194 Ga. App. 703, 391 S.E.2d 688 (1990)	§§ 13.22, 13.33
American Pac. Roofing Co. v. United States, 21 Cl. Ct. 265 (1990)	§ 9.5
Anuszewski v. Jurevic, 566 A.2d 742 (Me. 1989)	§ 13.33
Apex Control Sys., Inc. v. Alaska Mechanical, Inc., 776 P.2d 310 (Alaska 1989)	§ 9.9
Arbor Club v. Omega Constr. Co., 565 So. 2d 357 (Fla. Dist. Ct. App. 1990)	§ 14.7
Argentinis v. Gould, 23 Conn. App. 9, 579 A.2d 1078 (1990)	§ 13.16
Atheraton Condominium Apartment-Owners Ass'n v. Blume Dev. Co., 115 Wash. 2d 506, 799 P.2d 250 (1990)	§ 13.16
Bechtold Paving, Inc. v. City of Kenmare, 446 N.W.2d 19 (N.D. 1989)	§§ 9.2, 13.5, 13.17, 13.33
Bentley Constr. Dev. & Eng'g, Inc. v. All Phase Elec. & Maintenance, Inc., 562 So. 2d 800 (Fla. Dist. Ct. App. 1990)	§ 11.6
Bildoc, Inc. v. Chicago Hous. Auth., 714 F. Supp. 317 (N.D. Ill. 1989)	§ 10.29
Biomass One, Ltd. Partnership v. S-P Constr., 103 Or. App. 521, 799 P.2d 152 (1990)	§ 13.22
Birmingham, City of v. Cochrane Roofing & Metal Co., 547 So. 2d 1159 (Ala. 1989)	§§ 13.17, 13.23
Blau Mechanical Corp. v. City of N.Y., 158 A.D.2d 373, 551 N.Y.S.2d 228 (1990)	§§ 5.3, 9.2
Board of Supervisors v. Sentry Ins., 239 Va. 622, 391 S.E.2d 273 (1990)	§ 13.27
Bolton Corp. v. T.A. Loving Co., 94 N.C. App. 392, 380 S.E.2d 796 (1989)	§ 14.5
Cable Belt Converters, Inc. v. Alumina Partners, 717 F. Supp. 1021 (S.D.N.Y. 1989)	§ 9.5
Cam-Ful Indus., Inc. v. Fidelity & Deposit Co., 922 F.2d 156 (2d Cir. 1991)	§ 9.2

TABLES

Case	*Book §*
Care Sys., Inc. v. Laramee, 155 A.D.2d 770, 547 N.Y.S.2d 471 (1989)	§ 9.2
CCM Corp. v. United States, 20 Cl. Ct. 649 (1990)	§ 8.22
Century Constr. Co. v. United States, 22 Cl. Ct. 63 (1990)	§ 5.6
Charlie Brown Constr. v. Boulder City, 106 Nev. 497, 797 P.2d 946 (1990)	§ 13.12
Chas. A. Tompkins Co. v. Lumbermens Mut. Casualty Co., 732 F. Supp. 1368 (E.D. Va. 1990)	§ 13.27
Cigal v. Leader Dev. Corp., 408 Mass. 212, 557 N.E.2d 1119 (1990)	§ 13.16
Clements v. Barnes, 197 Ga. App. 120, 397 S.E.2d 560 (1990)	§§ 13.16, 13.33
Consolidated Fed. Corp. v. Cain, 195 Ga. App. 671, 394 S.E.2d 605 (1990)	§ 9.2
Corpus Christi, City of v. Heldenfels Bros., Inc., 802 S.W.2d 35 (Tex. Ct. App. 1990)	§ 11.6
Council of Unit Owners v. Freeman Assocs., 564 S.2d 357 (Del. Super. Ct. 1989)	§ 13.33
Credit Gen. Ins. Co. v. Atlas Asphalt, Inc., 304 Ark. 522, 803 S.W.2d 903 (1991)	§ 11.6
Criswell v. European Crossroads Shopping Center, Ltd., 792 S.W.2d 945 (Tex. 1990)	§ 12.14
Crown Oil & Wax Co. v. Glen Constr. Co., 320 Md. 546, 578 A.2d 1184 (1990)	§ 1.12
Crown Plastering Corp. v. Elite Assocs., Inc., 166 A.D.2d 495, 560 N.Y.S.2d 694 (1990)	§ 11.6
Dale R. Horning Co. v. Falconer Glass Indus., Inc., 730 F. Supp. 962 (S.D. Ind. 1990)	§ 5.3
David C. Olson Co. v. Denver & Rio Grande W.R.R., 789 P.2d 492 (Colo. Ct. App. 1990)	§ 5.6
DEC Elec., Inc. v. Raphael Constr. Corp., 558 So. 2d 427 (Fla. 1990)	§ 11.6
Deisch v. Jay, 790 P.2d 1273 (Wyo. 1990)	§§ 13.16, 13.32, 13.33
Desco Vitro Glaze v. Mechanical Constr. Corp., 159 A.D.2d 760, 552 N.Y.S.2d 185 (1990)	§ 10.2
Design & Production, Inc. v. United States, 18 Cl. Ct. 168 (1989)	§ 9.10
District of Columbia v. Campbell, 560 A.2d 1295 (D.C. 1990)	§ 13.27
D.M. Ward Constr. Co. v. Electric Corp., 15 Kan. App. 2d 114, 803 P.2d 593 (1990)	§ 1.12
Domingue v. H&S Constr. Co., 546 So. 2d 913 (La. Ct. App. 1989)	§ 13.19
Doughty v. Simpson, 190 Ga. App. 718, 380 S.E.2d 57 (1989)	§ 13.33
Douglass v. Liccardi Constr. Co., 386 Pa. Super. 292, 562 A.2d 913 (1989)	§ 13.32
Duncan v. Cannon, 204 Ill. App. 3d 160, 561 N.E.2d 1147 (1990)	§ 9.9
Earthbank, Inc. v. City of N.Y., 145 Misc. 2d 937, 549 N.Y.S.2d 314 (1989)	§§ 9.12, 10.30
Eastover Corp. v. Martin Builders, 543 So. 2d 1358 (La. Ct. App. 1989)	§ 13.16
Edwards v. United States, 19 Cl. Ct. 663 (1990)	§ 9.11
Elmira, City of v. Larry Walter, Inc., 150 A.D.2d 129, 546 N.Y.S.2d 183 (1989)	§§ 5.7, 14.5

CASES

Case	*Book §*
Employers Ins. v. Mississippi State Highway Comm'n, 575 So. 2d 999 (Miss. 1990)	§ 13.9
Fairbanks N. Star Borough v. Kandik Constr., Inc., 795 P.2d 793 (Alaska 1990)	§ 13.5
Fetzer v. Vishneski, 399 Pa. Super. 218, 582 A.2d 23 (1990)	§§ 13.16, 13.33
Fleming v. Urdl's Waterfall Creations, Inc., 549 So.2d 1057 (Fla. Dist. Ct. App. 1989)	§ 13.33
Floor Craft Floor Covering, Inc. v. Parma Community Gen. Hosp. Ass'n, 54 Ohio St. 3d 1, 560 N.E.2d 206 (1990)	§ 13.5
Florida Precast Concrete, Inc., *In re*, 112 B.R. 451 (Bankr. M.D. Fla. 1990)	§ 14.10
Fred A. Arnold, Inc. v. United States, 18 Cl. Ct. 1 (1989)	§§ 5.2, 14.10
Freeman v. Maple Point, Inc., 393 Pa. Super. 427, 574 A.2d 684 (1990)	§ 13.33
Fridlen v. Winchendon Hous. Auth., 28 Mass. App. Ct. 977, 553 N.E.2d 554 (1990)	§ 5.2
Garcia v. Kastner Farms, Inc., 789 S.W.2d 656 (Tex. Ct. App. 1990)	§§ 9.9, 13.32
Giuliani Contracting Co. v. United States, 21 Cl. Ct. 81 (1990)	§ 5.6
Glens Falls, City of v. Crandell Assocs. Architects, ___ A.D.2d ___, 566 N.Y.S.2d 689 (1991)	§ 13.19
Gorbett v. Claycamp, 553 N.E.2d 475 (Ind. 1990)	§§ 9.9, 9.13
Gunter Hotel v. Buck, 775 S.W.2d 689 (Tex. Ct. App. 1989)	§ 12.14
Guy Williams Realty, Inc. v. Shamrock Constr. Co., 564 So. 2d 689 (La. Ct. App. 1990)	§ 15.5
Gymco Constr. Co. v. Architectural Glass & Windows, Inc., 884 F.2d 1362 (11th Cir. 1989)	§§ 5.3, 9.10
Halterman v. United States Fidelity & Guar. Co., 269 Cal. Rptr. 363 (Ct. App. 1990)	§ 13.27
Henderson Inv. Corp. v. International Fidelity Ins. Co., 575 So. 2d 770 (Fla. Dist. Ct. App. 1991)	§ 1.12
Hermann v. Varco-Pruden Bldgs., 106 Nev. 569, 796 P.2d 590 (1990)	§ 13.17
Hernandez v. Westoak Realty & Inv., Inc., 771 S.W.2d 876 (Mo. Ct. App. 1989)	§§ 13.19, 13.32
Hieb v. Opp, 458 N.W.2d 797 (S.D. 1990)	§ 13.17
Hillcrest Country Club v. N.D. Judds Co., 236 Neb. 233, 461 N.W.2d 55 (1990)	§ 13.22
HOH Co. v. Travelers Indem. Co., 903 F.2d 8 (D.C. Cir. 1990)	§ 9.13
Houma, City of v. Municipal & Indus. Pipe Serv., 884 F.2d 886 (5th Cir. 1989)	§§ 13.5, 13.27
Howard v. Jay, 203 Ill. App. 3d 539, 561 N.E.2d 274 (1990)	§ 13.48
Husman, Inc. v. Triton Coal Co., 809 P.2d 796 (Wyo. 1991)	§ 8.22
Jacksonville, City of v. W.R. Fairchild Constr. Co., Ltd., 547 So. 2d 1010 (Fla. Dist. Ct. App. 1989)	§ 9.9
J.A. Jones Constr. Co. v. City of N.Y., 753 F. Supp. 497 (S.D.N.Y. 1990)	§ 9.14
Jerry B. Wilson Roofing & Painting, Inc. v. Jobco-E.R. Kelly Assocs., Inc., 151 A.D.2d 896, 542 N.Y.S.2d 867 (1989)	§ 10.35

TABLES

Case	Book §
J.M. Beeson Co. v. Sartori, 553 So. 2d 180 (Fla. Dist. Ct. App. 1989)	§ 14.10
Judd Supply Co. v. Merchants & Mfrs. Ins. Co., 448 N.W.2d 895 (Minn. Ct. App. 1989)	§ 13.27
Ken's Constr. Co. v. Liles, 560 So. 2d 103 (La. Ct. App. 1990)	§§ 9.12, 12.14
Kinney v. G.W. Lisk Co., 76 N.Y.2d 215, 556 N.E.2d 1090, 557 N.Y.S.2d 283 (1990)	§ 1.3
Knowles v. Westbrook Builders, Ltd., 188 Ill. App. 3d 343, 544 N.E.2d 121 (1989)	§ 13.32
Kohn v. Johnson, 565 So. 2d 165 (Ala. 1990)	§ 13.33
Korstad-Tebben, Inc. v. Pope Architects, Inc., 459 N.W.2d 565 (S.D. 1990)	§ 13.2
Krugman & Fox Constr. Corp. v. Elite Assocs., Inc., 167 A.D.2d 514, 562 N.Y.S.2d 188 (1990)	§ 13.47
Lakeview Constr. Co. v. United States, 21 Cl. Ct. 269 (1990)	§§ 5.6, 9.5
Lang Bros. Inc. v. United States, 20 Cl. Ct. 551 (1990)	§ 1.3
Lathan Co. v. United States, 20 Cl. Ct. 122 (1990)	§ 8.22
L.K. Comstock & Co. v. United Eng'rs & Constructors, Inc., 880 F.2d 219 (9th Cir. 1989)	§ 10.17
L. Loyer Constr. Co. v. City of Novi, 179 Mich. App. 781, 446 N.W.2d 364 (1989)	§ 9.5
Lochrane Eng'g, Inc. v. Willingham Realgrowth Inv. Fund, Ltd., 552 So. 2d 228 (Fla. Dist. Ct. App. 1989)	§ 13.33
Maitland Bros. Co. v. United States, 20 Cl. Ct. 53 (1990)	§ 9.11
Mancorp, Inc. v. Culpepper, 781 S.W.2d 618 (Tex. Ct. App. 1989)	§§ 13.19, 13.33
Manshul Constr. Corp. v. Board of Educ., 160 A.D.2d 643, 559 N.Y.S.2d 260 (1990)	§§ 5.2, 13.51
Markway Constr. Co. v. Kirchenbauer, 769 S.W.2d 836 (Mo. Ct. App. 1989)	§§ 9.6, 9.13
Marshall Constr., Ltd. v. Coastal Sheet Metal & Roofing, Inc., 569 So. 2d 845 (Fla. Dist. Ct. App. 1990)	§ 11.17
Martell Bros., Inc. v. Donbury, Inc., 577 A.2d 334 (Me. 1990)	§ 10.30
Mason v. Yontz, 102 N.C. App. 817, 403 S.E.2d 536 (1991)	§ 13.33
Massachusetts Bay Transp. Auth. v. United States, 21 Cl. Ct. 252 (1990)	§ 13.2
May v. Ralph L. Dickerson Constr. Co., 560 So. 2d 729 (Miss. 1990)	§ 13.19
McDevitt & Street Co. v. Marriott Corp., 713 F. Supp. 906 (E.D. Va. 1989)	§§ 9.11, 14.5
Mine Creek Contractors, Inc. v. Grandstaff, 300 Ark. 516, 780 S.W.2d 543 (1989)	§ 13.19
Mor-Wood Contractors, Inc. v. Ottinger, 205 Ill. App. 3d 132, 562 N.E.2d 1247 (1990)	§ 6.2
Mrozik Constr., Inc. v. Lovering Assocs., Inc., 461 N.W.2d 49 (Minn. Ct. App. 1990)	§ 1.3
National Sand, Inc. v. Nagel Constr., Inc., 182 Mich. App. 327, 451 N.W.2d 618 (1990)	§ 9.1

CASES

Case	*Book §*
Neal & Co. v. United States, 19 Cl. Ct. 463 (1990)	§ 9.15
Newark Beth Israel Medical Center v. Gruzen & Partners, 124 N.J. 357, 590 A.2d 1171 (1991)	§ 13.5
New York, City of v. Kalisch-Jarco, Inc., 161 A.D.2d 252, 554 N.Y.S.2d 900 (1990)	§ 13.22
Nohcra Communications, Inc. v. AM Communications, Inc., 909 F.2d 1007 (7th Cir. 1990)	§ 10.34
Norman Properties v. Bozeman, 557 So. 2d 1265 (Ala. 1990)	§§ 13.16, 13.33
North Star Contracting v. Long Island R.R., 723 F. Supp. 902 (E.D.N.Y. 1989)	§ 5.2
North W. Mich. Constr., Inc. v. Stroud, 185 Mich. App. 649, 462 N.W.2d 804 (1990)	§ 1.12
OBS Con. v. Pace Constr. Corp., 558 So. 2d 404 (Fla. 1990)	§ 11.6
Orndorff v. Chistiana Community Builders, 217 Cal. App. 3d 683, 266 Cal. Rptr. 193 (1990)	§ 13.33
Owners Realty Management Constr. Corp. v. Board of Educ. W. Islip, 160 A.D.2d 921, 554 N.Y.S.2d 648 (1990)	§ 9.9
Pawnee, Village of v. Azarelli Constr. Co., 183 Ill. App. 3d 998, 539 N.E.2d 895 (1989)	§ 13.17
Pensacola Executive House Condominium Ass'n v. Baskerville-Donovan Eng'rs, Inc., 556 So. 2d 850 (Fla. Dist. Ct. App. 1990)	§ 13.5
Placeway Constr. Corp. v. United States, 920 F.2d 903 (Fed. Cir. 1990)	§ 14.7
Pollitte v. Sherman, 168 A.D.2d 761, 563 N.Y.S.2d 915 (1990)	§§ 13.19, 13.24
Pools Mkts. S., Inc. v. Coggins, 195 Ga. App. 50, 392 S.E.2d 552 (1990)	§ 13.35
Port Chester Elec. Constr. Corp. v. HBE Corp., 894 F.2d 47 (2d Cir. 1990)	§§ 5.3, 9.1
Prairie Land Constr., Inc. v. Village of Modesto, 213 Ill. App. 3d 364, 571 N.E.2d 1210 (1991)	§ 9.11
Quate v. Caudle, 95 N.C. App. 80, 381 S.E.2d 842 (1989)	§ 13.33
Ramirez Co. v. Housing Auth., 777 S.W.2d 167 (Tex. Ct. App. 1989)	§ 10.29
Rapp Constr. Co. v. Jay Realty Co., 809 S.W.2d 490 (Tenn. Ct. App. 1991)	§ 13.12
Reliance Ins. Co. v. United States, 20 Cl. Ct. 715 (1990)	§§ 5.10, 9.13
Rembrant, Inc. *ex rel.* Wright Constr. Co. v. United States, 919 F.2d 1569 (Fed. Cir. 1990)	§ 9.9
Richardson v. Collier Bldg. Corp., 793 S.W.2d 366 (Mo. Ct. App. 1990)	§ 13.19
Richmond v. Grabowski, 781 P.2d 192 (Colo. Ct. App. 1989)	§ 13.9
Robert Irsay Co. v. United States Postal Serv., 21 Cl. Ct. 502 (1990)	§ 9.5
Ruby-Collins, Inc. v. City of Charlotte, 740 F. Supp. 1159 (W.D.N.C. 1990)	§ 8.3
Russo v. Heil Constr., Inc., 549 So. 2d 676 (Fla. Dist. Ct. App. 1989)	§ 14.13
R.W. Kern, Inc. v. Circle Indus. Corp., 158 A.D.2d 363, 551 N.Y.S.2d 218 (1990)	§ 13.22

TABLES

Case	*Book §*
Salard v. Jim Walter Homes, Inc., 563 So. 2d 1327 (La. Ct. App. 1990)	§ 13.33
Salvino Steel & Iron Works, Inc. v. Fletcher & Sons, Inc., 398 Pa. Super. 86, 580 A.2d 853 (1990)	§ 5.2
San Antonio, City of v. Forgy, 769 S.W.2d 293 (Tex. Ct. App. 1989)	§ 9.11
Sanchez Plumbing v. Aetna Casualty & Sur. Co., 564 So. 2d 1302 (La. Ct. App. 1990)	§ 11.17
Sarvis v. Maida, ___ A.D.2d ___ , 569 N.Y.S.2d 997 (1991)	§ 13.12
Schlothauer v. Gusse, 753 F. Supp. 414 (D. Mass. 1991)	§ 1.3
Servidone Constr. Corp. v. United States, 931 F.2d 860 (Fed. Cir. 1991)	§ 8.22
Shacocass, Inc. v. Arrington Constr. Co., 116 Idaho App. 460, 776 P.2d 469 (1989)	§ 9.11
Shook of W. Va. v. York City Sewer Auth., 756 F. Supp. 848 (M.D. Pa. 1991)	§ 14.5
Sorrels Steel Co. v. Great S.W. Corp., 906 F.2d 158 (5th Cir. 1990)	§§ 5.6, 9.2
Southeast Alaska Constr. Co. v. State Dep't of Transp., 791 P.2d 339 (Alaska 1990)	§ 14.10
Southeast Consultants, Inc. v. O'Pry, 199 Ga. App. 125, 404 S.E.2d 299 (1991)	§§ 13.5, 13.37
Spirit Leveling Contractors v. United States, 19 Cl. Ct. 84 (1989)	§ 9.5
State Dep't of Transp. v. American Ins. Co., 199 Ill. App. 3d 1068, 557 N.E.2d 932 (1990)	§ 11.14
State Highway Admin. v. Greiner Eng'g Sciences, Inc., 83 Md. App. 621, 577 A.2d 363 (1990)	§§ 5.2, 13.51
S&T Constr. Co. v. Harris, 789 P.2d 640 (Okla. Ct. App. 1989)	§ 13.9
Stewart v. C&C Excavating & Constr. Co., 877 F.2d 711 (8th Cir. 1989)	§ 11.17
Stone v. City of Arcola, 181 Ill. App. 3d 513, 536 N.E.2d 1329 (1989)	§§ 5.2, 5.8, 14.5
Sun-Cal, Inc. v. United States, 21 Cl. Ct. 31 (1990)	§ 10.28
Swain v. Harvest States Coops., 469 N.W.2d 571 (N.D. 1991)	§ 13.32
Thermoglaze, Inc. v. Morningside Gardens Co., 23 Conn. App. 741, 583 A.2d 1331 (1991)	§ 9.2
Thomas v. O'Brien, 791 S.W.2d 4 (Mo. Ct. App. 1990)	§ 10.35
Thomas Crimmins Contracting Co. v. City of N.Y., 74 N.Y.2d 166, 542 N.E.2d 1097, 544 N.Y.S.2d 580 (1989)	§ 9.11
Tilcon Gammino, Inc. v. Commercial Assocs., 570 A.2d 1102 (R.I. 1990)	§ 9.3
Timberland Paving & Constr. Co. v. United States, 18 Cl. Ct. 129 (1989)	§§ 10.40, 14.5
Transdulles Centre Ltd. Partnership v. USX Corp., 761 F. Supp. 430 (E.D. Va. 1991)	§ 13.27
Transpower Constructors v. Grand River Dam Auth., 905 F.2d 1413 (10th Cir. 1990)	§ 7.13
Travelers Indem. Co. v. Hennepin County, 918 F.2d 66 (8th Cir. 1990)	§ 13.27
Troise v. United States, 21 Cl. Ct. 48 (1990)	§ 14.5
Troyer v. Webster Homes, Inc., 566 So. 2d 114 (La. Ct. App. 1990)	§ 13.24
Turner Brooks v. Bowling Green State Univ., 51 Ohio Misc. 2d 1, 554 N.E.2d 956 (Ct. Cl. 1989)	§ 8.15

CASES

Case	*Book §*
Turner Constr., Inc. v. American States Ins. Co., 397 Pa. Super. 29, 579 A.2d 915 (1990)	§ 13.19
Uhley v. Tapio Constr. Co., 573 So. 2d 390 (Fla. Dist. Ct. App. 1991)	§ 13.19
Union Ins. Co. v. Bailey, 234 Neb. 257, 450 N.W.2d 661 (1990)	§ 12.7
Unis v. JTS Constructors/Managers, Inc., 541 So. 2d 278 (La. Ct. App. 1989)	§ 11.16
United States *ex rel.* Control Sys., Inc. v. Arundel Corp., 896 F.2d 143 (5th Cir. 1990)	§ 5.2
United States *ex rel.* Pertun Constr. Co. v. Harvesters Group, 918 F.2d 915 (11th Cir. 1990)	§ 5.2
Vaccaro v. Smith, 29 Ark. App. 175, 779 S.W.2d 193 (1989)	§ 12.7
Walsky Constr. Co. v. United States, 20 Cl. Ct. 317 (1990)	§ 5.6
Watson, Watson, Rutland v. Board of Educ., 559 So. 2d 168 (Ala. 1990)	§ 13.5
Weaver-Bailey Contractors, Inc. v. United States, 19 Cl. Ct. 474 (1990)	§§ 9.13, 14.5
Wenzel v. Boyles Galvanizing Co., 920 F.2d 778 (11th Cir. 1991)	§ 1.3
Westates Constr. Co. v. City of Cheyenne, 775 P.2d 502 (Wyo. 1989)	§ 9.5
Western Empire Constructors, Inc. v. United States, 20 Cl. Ct. 668 (1990)	§ 9.9
Western Insulation Servs., Inc. v. Central Nat'l Ins. Co., 460 N.W.2d 355 (Minn. Ct. App. 1990)	§ 13.27
White Oak Corp. v. Department of Transp., 217 Conn. 281, 585 A.2d 1199 (1991)	§§ 5.2, 13.51
Wilhite v. Brownsville Concrete Co., 798 S.W.2d 772 (Tenn. Ct. App. 1990)	§ 13.33
Winford v. Webster Gravel & Asphalt, Inc., 571 So. 2d 802 (La. Ct. App. 1990)	§ 13.17
Woodhaven Homes, Inc. v. Kennedy Sheet Metal Co., 304 Ark. 415, 803 S.W.2d 508 (1991)	§ 9.5
Woodruff v. Johnson, 560 So. 2d 1040 (Ala. 1990)	§ 13.9
Wright Way Constr. Co. v. Harlingen Mall Co., 799 S.W.2d 415 (Tex. Ct. App. 1990)	§ 13.16

INDEX

AGENTS
    Authority of §§ 5.3, 9.10

BONDS
    Requirements for public works projects § 13.27

CAUSES OF ACTION AND RECOVERIES
    Supplier's defective performance § 5.3

CHANGES IN SCOPE CLAIMS
    Liability of accepting authority §§ 9.2, 9.13

DEFENSES
    Acceptance by owner or owner's agent §§ 9.2, 9.13, 13.17, 13.23

PAYMENT DELAY CLAIMS
    Construction of contract clauses § 11.6

SUBCONTRACTORS
    Breach of contract regarding payment of § 10.17
    Rights on public works projects § 13.27

SUPPLIERS
    Rights on public works projects § 13.27

TERMINATION CLAIMS
    Acceptance of turnkey project § 10.29

ISBN 0-471-55397-2